Ranching
Like a 12-Year-Old

Ranching
Like a 12-Year-Old

Ranching that is simple, easy, and fun!

Tom Krawiec

RANCHING LIKE A 12 YEAR OLD
Ranching that is Simple, Easy, and Fun

Copyright © Tom Krawiec, 2022

Published by Tom Krawiec, Athabasca, Canada

Edited by Kiley Venables
Structural edit by Dale Youngman

ISBN:
    Paperback   978-1-77354-424-3
        ebook   978-1-77354-425-0

Publication assistance and digital printing in Canada by

PAGEMASTER
PUBLISHING
PageMaster.ca

Foreword

It was over 20 years ago when I first ran into Tom Krawiec. Back then, he was a young man with a spring in his step and a twinkle in his eye. You would have to look long and hard to find anyone else who was as passionate about grazing and ranching as Tom. His desire to keep learning, his outside the box thinking and his innovative ideas have been so inspiring to me over the years.

He and I have spent countless hours debating, brainstorming, and sharing ideas back and forth about everything grazing. I definitely can't leave out the hours we spent discussing all of the mistakes we've made along this journey. I have been blessed and consider it an honor to call him a friend.

Today, that passion in his eye still shines through every time we talk. He has been dedicated to this industry and remains a huge advocate to improved grazing management. His years of experience, trials and experiments makes him a great resource to anyone looking to improve their management on the ranch. I would like to give a huge thank you to Tom for always being there for me, to challenge me, to push me and to inspire me in this thing we call Life.

Tom is one of my go-to guys when it comes to grazing, there is so much knowledge to be gained through this book that you now hold in your hands!

To quote Aristotle, "The more you know, the more you know you don't know." I hope you enjoy Toms experiences and wisdom as much as I have over the years. Never stop learning.

God bless you and your family
Steve Kenyon
Greener Pastures Ranching Ltd.

Contents

Introduction

B y the title of this book, you might assume I wrote it for beginner ranchers. Let me assure you that this is not a "How to Become a Rancher" essay. Instead, I will share my thoughts and experiences on how to set up a ranching operation so that it is simple, easy, and fun. It is my firm belief that if ranching is not simple, easy, and fun, we are doing something wrong.

Many people will disagree with my methods, and that is fine: I certainly don't have a monopoly on how things should be done! I will state, though, that I abhor tedious work. My goal has been to avoid tedious work as often as possible. When it comes to tedious work, I have found some way to simplify it or avoid it altogether. I have had many heated discussions with people who get great satisfaction from working hard. Our thinking just does not mesh.

I would not say I am lazy. In fact, when I set up a new ranch or start a new venture, I work my ass off for long hours. However, I am always thinking of ways to lessen the labour or the outlay of cash. I want things set up so it is a one-person show, and that one person can be low skill and/or low energy.

Throughout this book, there may be ways of doing something that don't make sense to you right away. Please trust that everything I do is to save either labour or money. Also, let me point out that I'm not very handy when it comes to repairing or building things, and I do mean Not Very!

If you are a person who wears your work ethic as a badge of honour, then you might not get much out of a book about how to avoid work. However, if you want to figure out how to decrease your workload, then keep on reading.

Background

Here is some background on how I came to think the way I do. In January 2000, before we began farming, my wife and I attended a Holistic Resource Management (HRM) course. That summer we started grazing hogs, cattle, laying hens, turkeys, and horses. At the course, we were taught to keep our costs as low as possible. Following this thinking, we didn't build any permanent fence or water systems; we built temporary paddocks and hauled water daily. Although we were in awesome shape, the labour involved was crazy!

In 2001, my friend's twelve-year-old daughter Heather and her thirteen-year-old friend Tiffany came to stay with us for the summer. Since our system was so labour intensive, it was very difficult for the two girls to help in a meaningful way. By the end of our second season, I realized we needed to do things differently.

Heather and Tiffany came for the next several summers. Each summer, we had to simplify our daily chores. Everything had to be constructed so that the girls could handle it, which made the chores easy for us as well.

The first thing we simplified was water delivery. Our HRM instructor showed us the old water truck he used to haul water to his cow herd. He and his HRM friends competed to have the oldest truck to haul water. Following suit, we built a deck on the frame of an old hay rack and mounted a 1250-gallon plastic tank on it. This was quite labour intensive to use because the wagon had to be brought up to the house, filled, then hauled back to the herd. The float used to fill the water trough was home-made (another piece of HRM advice) and frequently leaked, which created a mud hole and caused us to refill the tank often. Something had to be done to improve our water system.

The provincial government offered a grant for water pipeline, so I planned out a pipeline system to distribute water on the entire 373 acres we owned. At the time, 1¼-inch polyethylene pipe was all we required to supply water. This did work great for several years, until we started to expand.

The next project we tackled was permanent fencing. We started out hand-pounding ¾-inch rebar posts, then stringing high tensile wire. At the time, rebar was cheaper than wood posts and we didn't require a post pounder. However, there were always issues with the wire grounding out on the steel posts, which created more labour. After a few years I realized wood posts were the way to go.

I share our early years only to demonstrate that we tried a number of things before we arrived at the template I now use and recommend. However, it is really a way of thinking that has evolved to get to a system where a twelve-year-old or an eighty-year-old can ranch effectively.

This book is divided into six sections: Grass Management, Water, Fencing, Animal Handling, Economics, and Business. Each section is important to make the whole system work. Ranching cannot be simple, easy, and fun if one of the parts is out of line. These are my ideas on how to make an operation scalable.

That means, if you add another hundred animals, neither your labour nor your infrastructure cost will increase. Once you have a scalable system, the sky's the limit!

Most of my ideas in this book have come from someone else. As well, a lot of my thinking developed after attending the Ranching for Profit School and the follow-up program Executive Link. I may have put my own twist on things, but there isn't much that is an original idea. If I can remember where an idea came from, I will cite the source. Often, though, I don't remember where an idea came from because I have been integrating it into my work for years.

Before I continue, my editor told me to introduce myself, even though my name is on the cover!

Since you already know my name, let me give you my history.

My parents owned an oilfield trucking company, which means I grew up in an oil town and not a ranching town. However, since both my parents grew up on farms, many of our relatives farmed. My heart was in farming, but agriculture was only a summer vacation for me when I was young. Although I imagined buying my grandparents' place when they retired, for a twelve-year-old kid, it was just a pipe dream.

My paid work experience consisted of several jobs related to the oil business. I drove trucks hauling water, oil, and gravel. One water hauling job led me into the world of drilling rigs, and I met one of my most influential teachers, Duane Carol. I worked my way up to driller and after three years of that, I finally realized I really wanted to be farming. That's when my wife and I took my oilfield savings and bought the in-laws' place. In January 2000 we took the Holistic Resource Management (HRM) course. That set my course for lifelong learning, developed my passion for grazing, and transitioned us into ranching. On this journey, the Ranching for Profit School has been an enormous help in developing my business acumen. That's enough about me, so let's jump in!

Oh, one last thing before you start, please remember that 83 percent of all statistics are made up on the spot and not everything you read is completely true! Enjoy.

CHAPTER 1

Can a 12yr Old Do This?

In the spring of 2022 the online magazine On Pasture published a number of articles discussing how to balance life with work. One of the authors, Troy Bishop, wrote a few articles on how to fit grazing around your life rather than building your life around your grazing operation. Troy wrote about planning camping trips, vacations, & time with his grandkids. I think he was spot on in his thinking. It is very common in agriculture to forget that there is life outside of ranching. As a person who grew up with parents who owned a trucking company, I dare say being consumed by your business occurs outside of agriculture as well.

If you followed Troy's articles, then you know he puts a great deal of emphasis on the grazing chart. Without a plan that the grazing chart offers, it is pretty difficult to ensure you make time for a life outside of grazing. The question then becomes, 'How do I get there?' Just because you plan to vacation from July 1 to July 8 doesn't mean the logistics of your operation will allow you to take that time away. Of course, you could just take the time, but personally I would rather take time off and come back to an operation that is still running smoothly.

Early in my ranching journey, I realized labour was the biggest factor limiting growth and enjoyment of life. Sure I enjoyed ranching, but that was all I was doing. Further, I was going to bed every night exhausted! There was no time for anything except work. Thank heavens for Heather & Tiffany.

Heather is the daughter of friends who lived in town, and Tiffany was Heather's best friend. The two girls started coming for the summer when they were 12 and 13. They were goofy, creative, up for anything, and hardworking. The three of us had a blast together!

Heather & Tiffany taught me that for ranching to work for them, everything had to be set up so they could do it. Gates had to be easy to open. Paddock moves had to be simple. Equipment had to be light. Basically, things had to be set up so a 12yr old could do it. It was those summers that got me questioning everything I did. To this day, every time I have work to do, I ask myself, "Could a 12yr old do this?" If the answer is no, then I either change it or I stop doing it.

The cool thing about having your operation simple enough for a 12yr old to run is that an 80yr old can run it as well.

Simplify

Once I started making changes, it only took two years to get to the point where the two girls could run the day-to-day operations of the ranch. In fact, our family went on a week-long horse trip to the mountains and left the two girls to run the ranch. We hadn't expanded yet, but the girls still had 80 cow/calf combo's (their words not mine), 50 pastured hogs, and 30 ewes with lambs to look after.

It was a joy to come home and find everything in order. It was an even bigger joy to see the pride Heather & Tiffany had for having done a great job while we were gone. Certainly those two girls were exceptional, but what really made their success possible was that the operation was set up to be simple, easy, and fun.

It is my firm belief that if you routinely ask yourself the question, "Can a 12yr old do this?" in two or three years, you will have the time to enjoy life outside of ranching. In Ranching For Profit, we were taught the difference between working in the business (WITB) versus working on the business (WOTB).

Working In The Business is the day-to-day operations of running a ranch. Working On The Business involves goal setting, planning, creating a vision for your life, etc. To get to WOTB work, the WITB work needs to be reduced. In my experience, the question "Can a 12yr old do this?" is a very effective tool to get yourself to a place where WOTB is easy to incorporate into your day.

My friend Steve Kenyon talks about a time he went to Columbia (the country) for a wedding. His hired hand was not capable of running things for the 10 days Steve would be gone. Steve was about to cancel his trip when he had an epiphany. He could just open up several paddocks at once and let the animals have access to all that grass. He accepted that he would be violating the graze period and probably mess up his recovery time, but he would be able to attend his friend's wedding.

There are times when the decision to get away is a must do (medical emergency, mental health day, death, etc). However, when it comes to grass management, this decision should not be taken lightly.

In my experience, when managing grass, you can not violate the concepts of Recovery Period and Graze Period. These two concepts drive grass production, which in turn drives the economic engine of every grass based operation. As Troy points out in his Holistic Goal Setting, economics is one of the pillars you need to consider.

A key component of economics is profit. To create profit your pastures must perform really well and to do this you need to graze in the sweet spot. (Don't worry! I will be explaining the 'Sweet Spot' in depth shortly)! Grazing in the sweet spot does not allow for recovery periods that are too long in the same way it does not allow for recovery periods that are too short. It has been my experience that most forages grow back at less than 50% of possible production if they are grazed after they reach reproductive stage. Once again, decreased production affects the economic driver of your operation. To counter decreased production, maintaining both the appropriate recovery period and the graze period must be adhered to.

The graze period is the time animals are in a paddock. At the latitude I have done most of my grazing, the graze period during May/June is three days. After the first week of August, that period is normally 5-7 days. For my planning, August is the best time to take a vacation.

In June it may not seem like a big deal to leave animals in a paddock for five days so you can take an extended weekend. However, when we violate either of the two concepts the damage is done underground where we can't see what is happening. If we caused damage to our animals we could easily see what happened.

It is difficult to see when we cause damage to plants and soil biology.

Troy and others have espoused the value of using a grazing chart to plan for life outside of ranching. I whole heartedly agree and have been using a grazing chart to help plan my life for over twenty years. I would be lost without it.

Results

The two ideas I have shared help me maintain a much better balance between life and work. It may seem corny, but by asking the question, "Can a 12yr old do this?" I was able to set up a custom grazing operation of 3,000 yearlings that only took six hours of work per day. Once that state was achieved, we had fun-filled summers of lake days, poker nights with neighbours, fishing trips, personal development courses, laying in the grass, etc. The funny thing is, it wasn't difficult to get there. It just took some conscious intention and then the motivation to make changes.

As far as my points on grass management, I may be too ridged. It has been my experience, though, that if I do not stay ridged, production suffers. That is why when planning, I am always cognizant of Recovery Period and Graze Period.

As Troy points out in his articles, life is too short and too fickle to work it away. What I have shared is how I approach grazing so that I have a life outside of work.

CHAPTER 2

Grass Management

As I mentioned in the introduction, we took an HRM course in 2000. It was there where we were introduced to Management Intensive Grazing, which is now commonly called Planned Grazing. My interest in grass management was piqued right away, but once I read *Grass Productivity: An Introduction to Rational Grazing* by André Voisin, interest turned into passion!

To me, *Grass Productivity* is the bible for any grass manager. I thought everyone who had anything to do with grass would devour that book in the same way I did. As it turns out, the only person I know who has read it and feels so passionately about the book... is me.

Grass has taken hold of my entire being. My thinking is consumed by growing more grass, stockpiling more grass, and improving the nutritional quality of grass. Since year four, I have been able to feel the energy coming from the grass. I know this sounds hokey but when I walk through a paddock I either feel excited, depressed, or calm depending on how the grass is growing.

I have learned some costly lessons because I disregarded animal performance, cash flow, or emotional balance for our family. I learned through the School of Hard Knocks that there is more to ranching than growing great grass. I have been putting more effort into learning those other skills ever since. However, the system still has to begin with the grass. Period!

To get a good handle on the concepts I will discuss in this section, I am putting the glossary at the start rather than in the appendix. One other note, the examples I use are based on the area I have been grazing in for 20+ years, specifically the latitude. I make that point because latitude determines how much sunlight you receive in a given day. There is more daylight during May and June at latitude 54* then there is at latitude 51*. This is to say that you may have to adjust my examples slightly to fit with the daylight hours at your latitude.

Glossary

Standard Animal Unit (SAU): 1SAU is the value for one 1000lb dry pregnant cow. The value increases or decreases depending on the class of animal. This system was developed from research by A. H. Penderis and can be found in a table at the back of the *Aide Memoire for Holistic Grazing Planning*.

Example: a 1320lb dry pregnant cow has a value of 1.23SAU

Stock Day (SD): the amount of forage an animal eats in one day. 1SD is the amount of forage one 1000lb dry pregnant cow consumes in one day. The value increases or decreases based on the class of animal you are grazing. When dealing with multiple animals you multiply the value of each animal by the number of animals in the group. 1SD is the equivalent of 24lbs of dry matter.

Example: one 1320lb dry pregnant cow has a value of 1.23SD. If you have a herd of 150 cows, the herd's value is 184SD's (150 animals X 1.23SAU = 184SD)

Stock Days Per Acre (SDA): the amount of forage consumed/harvested from an acre of land in one day. I like using SDA's because it gives you a production figure for each acre of land. It is similar to a grain farmer calculating how many bushels of grain are produced per acre. This figure can easily be converted to revenue per acre.

Example: if your paddock is 10 acres and you graze a herd of
150 dry pregnant cows for three days, you harvested 55SDA
((150 cows X 1.23SAU X 3 days)/ 10acres = 55SDA).

Paddock: a fenced area where animals graze; designed and subdivided for rotation or limited turnout. They are part of a pasture or pasture system.

Recovery Period: the amount of time a paddock is left to regrow with no animals grazing. It is calculated from the day animals leave the paddock until the day they return.

Example: if your animals are moved out of a paddock on
May 25 and return on June 29, the recovery period is from
May 25 to June 28 or 35days.o

Graze Period: the amount of time your animals are left in a paddock. The animals should be removed from a paddock before they are able to take a bite of new growth. You can determine the max graze period by strip grazing a paddock where the animals have to walk back for water. The animals will stay in the new strip of grass you give them until there is new growth in the first strip. Once there is enough new growth in the first strip, they will stay there rather than going to the next strip you give them.

Example: pick a paddock that will last 6 days and set up
your first strip. Leave the water in the first strip. Give the
animals a new strip each day. They will walk to water in
the first strip and return to the strip you just gave them
until there is enough new growth in the first paddock. If
the animals keep returning to the last strip, then you know
your max graze period is more than six days. However, if

*on day four, there are animals staying in the first strip, then
you know your graze period is three days because by day four
there is enough new growth for them to graze. The animals
should be removed before they can graze new growth.*

Why 13 Paddocks: 13 paddocks is the minimum number of
paddocks I recommend for each group of animals. Thirteen
paddocks will allow you to respect both the recovery period and
the graze period and gives you a bit of flexibility to compen-
sate for different forage production between paddocks. Both
recovery period and graze period must be respected to be a
successful grazier. *How to determine Graze Period is discussed on
pages 37-38*

Stages of Grass Growth: there are three major stages of
grass growth.

Stage One: characterized by slow, highly palatable growth. The
type of growth found in early spring.

Stage Two: grass grows rapidly because of increased leaf area
(solar panel) and is also very palatable.

Stage Three: grass grows slowly, is setting seed, and is not very
palatable nor nutritious.

**for an in-depth explanation of these stages, watch 'Grass Growth
Stages and Grazing Impacts' from the University of Wyoming.*

The Sweet Spot: a stage of growth just before Stage Three.
It refers to the whole paddock and not individual plants. The
sweet spot is when 15-20% of the plants in a paddock are in
Stage Three.

Sward: the forage within a paddock

TDN: (Total Digestible Nutrients) the digestible portion of forage. It is directly related to energy in the feed.

The Grass

CHAPTER 3

Grazing in the Sweet Spot

Grazing in the Sweet Spot is a phrase I have been using for over a decade. Until the last few years, I have only used that term when talking to myself. I didn't have a good handle on how to explain it to others. Now that I have thought about it for a number of years, I feel I am ready. That explanation is what this section is about.

Let me first explain why I graze in the sweet spot. First off, grazing in this zone is the fastest growth of the sward. If we keep the sward in the sweet spot, we will realize much better forage production than grazing on either side of the sweet spot. The more forage we produce, the more animals we can graze or the longer we can graze them.

Another benefit of grazing in the sweet spot is increased animal performance. Forage at this stage is neither too moribund nor too washy for animals to experience good gains. It is common to realize 50-80lbs increased weaning weights on calves when they have been grazing in that zone all summer. In the years I have grazed sheep, I have sold 115lbs lambs that are only 4 ½ months old when grazing in the sweet spot.

Keeping your sward in the sweet spot also promotes increased forage production. Dr. Kris Nichols speaks about how soil biology needs to be fed more than once a year to be healthy and robust. When plants are kept in a vegetative state such as the sweet spot, they release exudates into the soil to feed the biology. Once the plant goes past the sweet spot, or into the reproductive phase, the exudates go into the seed head preparing for reproduction. Once that happens, the biology goes to sleep. When soil biology is dormant, it is pretty difficult for forages to grow. As you know, soil biology is the driving force of a grazing system. Therefore, it should be our goal to get the soil biology as healthy and productive as possible.

Now for the big question: What is the Sweet Spot? To answer that question, you must first understand the growth curve of plants.

There are three major phases of grass growth. The first stage is characterized by slow, highly palatable growth. It is what you experience in the spring. Stage-Two grass grows rapidly because of increased leaf area (solar panel) and it is also very palatable. Stage-Three grass grows slowly, is setting seed, and is not very palatable nor nutritious. The Sweet Spot is in late Stage Two growth just before going into Stage Three. However, there is a caveat; the Sweet Spot relates to the sward as a whole and not individual plants.

Since species of plants grow at different rates, it is next to impossible to graze at a time when all plants are in late Stage Two. Some species like timothy and smooth brome get into Stage

Three very rapidly. That means when you are grazing in the sweet spot the sward will have some plants in the reproductive phase. It does not mean all plants will be in Stage Three. A certain publisher chastised me for not explaining the three stages of grass growth. Since there is a great video in the On Pasture magazine archives explaining the three phases of grass growth, I will not do it here (see Grass Growth Stages and Grazing Impacts' from the University of Wyoming). Instead of detailing the growth curve of individual plants, let me explain how I determine the stage of growth for the sward.

Like I stated earlier, there will be some plants that are in Stage Three growth. Therefore when I walk through a paddock, I don't pay attention to the timothy or smooth brome. I look at the other forages. Are the other forages starting to flower or get into the boot stage? If 10-15% of the other forages are in the boot stage then for me that means the sward is in the sweet spot. Of course I could be wrong, so I use another indicator to check my judgment. Animal dung tells me if I'm grazing where I want to be.

Animal dung will show whether or not your animals are grazing the sweet spot. For cattle this means a cow pat that is 2" high. If the pat is closer to 4-5" then you know you are grazing in Stage Three. If there is not really any pat, but more like lava, then you know you are too early and the sward needs more time to recover.

With sheep dung, you know you are in the sweet spot when it is mushy and not pellets.

When grazing horses, you will be in the sweet spot when the dung resembles a cow pat that stands 4-5" high. If the dung is splashy you are grazing too early. Dodging splashy horse dung can be a real hazard when driving a carriage; just saying!

Grazing in Stage 3   *Grazing in the sweet spot*   *Lava*

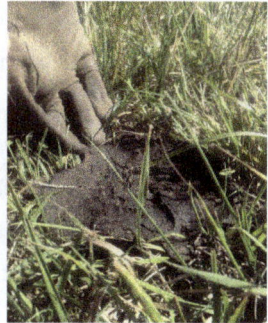

Grazing the sweet spot is also how I consistently have high quality, high-volume stock piled forage. When you have high quality, high-volume stock piled forage, the economic benefit is huge. Every day you graze instead of feed your animals is money that stays in your pocket. If you are a custom grazier, every day you can keep the herd means more money that goes into your pocket.

Grazing in the sweet spot is like a triple threat. You benefit by increased animal performance, improved forage production, and you benefit economically. I have been grazing in the sweet spot for over 10yrs and each year it warms the cockles of my heart to see the improvement in these three areas of the grass I manage. I know there are other philosophies of grazing, but based on my experience, grazing the sweet spot consistently shines in dry years, wet years, and all the years in between.

I will now share some science that supports what I have been doing intuitively for the last 12-15 years. Further, I will challenge the notion of long recovery periods based on science and my own observations.

As I have mentioned, Dr Kris Nichols, a soil biologist from Minnesota, explains that for soil biology to be highly active, we need to feed the soil more than once a year. When we graze swards with long recovery periods, the exudates normally released to the soil by vegetative plants, instead go into the formation of seeds. When this happens, Dr Nichols posits that the biology in the soil goes into semi-dormancy. Therefore, as graziers, we must keep forage in a vegetative state to enhance life in the soil.

Research done by Aulakh et al in 2001, supports what Dr Nichols suggests. In that research they found "seedlings produce the lowest amounts of root exudates; this gradually increases until flowering and decreases again at maturity." These findings were supported by work from Garcia et al. (2001) where they found "that root exudation is positively correlated with root growth; it means that actively growing root systems secrete more exudates."

If you can accept that soil life is nourished by root exudates, then the case can be made that the more we feed the biology in the soil, the healthier they will become. They will be more active, more robust, be more resilient to pathogens, and reproduce more. All these attributes contribute to a healthy, thriving, and increasing population.

Most of the support I have for the benefit of grazing in the sweet spot is based on my own observations, which are certainly open to criticism. I am not a scientist, so my findings are not based on rigorous scrutiny. However, I do have production records which record how much forage was harvested from each paddock. The metric I use here is Stock Days Per Acre (SDA). One SDA is the amount of forage one 1000lb dry pregnant cow consumes in one day. I use stock days per acre to assess whether a grazing practise is working.

By using SDA's, I have discovered that when a sward is given an overly long recovery period, it grows back to 50% or less of what it would produce if grazed in the sweet spot. As a producer, that is very important if you make your living off your grass. The sneaky thing about this scenario is that you don't know that you missed out on a pile of forage because it is difficult to evaluate something that isn't there.

The only reason I know this is because I have kept records on my grazing charts for 20+ years and have seen a pattern emerge in relation to recovery periods. In fact, when people lament about the lack of grazing they achieved in a season, it is invariably traced back to letting forage get into reproductive phase. They either started too late or their recovery period was too long.

If the goal for a grazier is to keep your sward vegetative, then you have to start grazing before all your paddocks are ready. Jim Gerrish explains this very well in his book MiG so I will not regurgitate his explanation here. Suffice it to say that starting early is imperative to grazing in the sweet spot. Let me share an example of how this practise helped me in 2021.

The summer of 2021 I was managing the grass for about 2,400 head of cattle (pairs, bulls, breeding heifers, and grass finishing heifers). The conditions started out great. We had good moisture from 2020 and the temperature was increasing nicely. However, Environment Canada was forecasting a severe drought.

Instead of getting scared by the forecast, I pushed to get cattle out grazing the second week of May. My goal was to keep forage in the sweet spot and I knew if I didn't get out when I did, the cattle would be grazing old forage by mid-July.

For this latitude, Athabasca, Alberta, based on my experience, the recovery period in May – June is 35 +/- 5 days and the graze period is 3 days max. Using this information, I filled out my grazing plan.

The plan worked well and all but two herds finished their first rotation by June 21. This date is important because it seems forage has an internal clock that drives plants to reach reproduction by the longest day of the year, which is June 21. (Again, this is my observation and not a scientifically researched phenomenon). As a grazier, it is my goal to hamper plants from reaching maturity so I had better be through my rotation by June 21. As I pointed out, my mission was accomplished in 2021.

Conditions changed late in June and after the first rotation, I was confronted with extreme heat and lack of moisture. This was the first time I was confronted with these two conditions at the same time for an extended period.

The grass growth slowed right down to the point where I thought it was dormant. It wasn't actually dormant, but regrowth was severely hampered. So much so the graze period went from three days to five days for the month of July. I slowed down the rotation accordingly. Instead of using 35 days as my target, I started using 50 days as the recovery period.

We received ½" of rain the last week of July and the temperature started to drop to normal. Then something amazing happened the first week of August; the grass started growing faster than I have ever seen at that time of year. To me it seemed as though the grass was exhibiting compensatory growth to make up for its lack of growth in July. Once again, I sped up the rotation. In simple terms, I just shortened the graze period.

I was quite nervous about the results to expect, but once again, by keeping grass in the sweet spot, I was rewarded with an extended grazing season. The rancher I worked for grazed until the end of November when most producers in the county were off grass October 1. He also had about twenty days of high quality stockpiled forage for 1000 cows to calve on. For that producer the extra grazing he received in a year where hay was $220/ton was a significant cost saving. For his business, reducing winter feed cost is what he is most concerned about.

I have been challenged numerous times that long recovery periods promote diversity in your pastures, and I have to agree with that statement. However, grazing in the sweet spot also promotes diversity. Does one system promote more diversity than the other? I don't know and neither do the people who make this claim. What I do know, though, is that increased production is achieved when grazing in the sweet spot. The records I keep prove that. I have also recorded the forage produced when I let forage get into reproductive phase and when I kept it in the sweet spot. Grazing in the sweet spot produces more forage and better gains year after year in wet and dry conditions.

Let me close off this section with a bit more science. I have been told by many people that planned grazing increases carbon stored in the soil. For many years I too believed that. Unfortunately, research does not support that claim.

To investigate this notion, a group of scientists from the University of Alberta developed a five year-long study to test this theory (Bharat et al 2020). Surprisingly to most people, including the researchers, there was no difference in stored carbon between what they called Adaptive Multi-Paddock Grazing (planned grazing) and continuous grazing at a moderate level.

These findings support the conclusions found in many other studies.

In a review of the research investigating non-selective grazing, also called planned grazing or the Savory method, by M. Nordborg in 2016, found no benefit compared to moderately-stocked continuous grazing. I am not surprised by the findings when I read how the studies were designed.

In each study, what I see is that the recovery period is too long, the graze period is too long, or grazing started too late in the spring. Whatever the case, the end result was the same; forage was allowed to get into the reproduction phase.

In the case of Bharat et al 2020, the study's premise was that early season grazing was anything before July 31. It has been my experience that where most of the data was collected, if the grass has not been clipped by June 30, it has reached maturity. By not clipping grass before June 30, it means all paddocks will be in re-productive phase.

As you read earlier, Aulakh et al (2001) demonstrated in their research that once a plant reaches maturity, it basically stops feeding the soil biology. When the biology is not being fed, it goes into a dormant state as suggested by Dr Nichols. This does not promote a robust system in the soil. A healthy community below the ground is what drives a robust plant population above the soil.

I know I have poked several bears by suggesting that too long of a recovery period is as bad as not enough recovery. That is not my intent. My goal is to point out that we can do better as graziers and we should. We compete for land with a highly subsidized grain business, and there is the belief that grazing can't produce what a crop can produce. This belief is so widespread that it is taken as fact. I don't buy it and I hope more people start feeling the same way.

In my own journey as a grazier, the best production I have done is 525 SDA from one paddock in one year. That is the equivalent of 12,600 lbs dry matter (DM) per acre.*

On numerous occasions, I have achieved 325 SDA (7,800lbs DM per acre). Converted to $/ac, based on custom grazing rates in our area, those paddocks produced $372.75/ac and $230.75/ac respectively. It is much easier to compete for land when your land is producing those kinds of returns.

It is my contention and experience that the way to get that kind of production is by grazing in the sweet spot. There is no science to back up that claim other than my own production records. As a producer, though, isn't production from our pastures really what we are after?

*1 stock day per acre (SDA) is equivalent to 24lbs of dry matter

CHAPTER 4

Graze Planning

I t all starts with the grass! If you do not get a handle on the grass, the rest of your system will not work.

There are many ways to prop up your grass. Depending on who is selling, it can be in the form of mechanical, chemical, or seed amendments. Magazines are full of ads detailing the amazing results of seed varieties, the latest chemical weed suppressant, or mechanical stimulation. However, there are no ads for the most effective method; it can't be bought from your local sales rep. You do not need to know what soil type you have or what bugs are in your soil. The formula is something you can get for free and will last longer than any input you can purchase. This method is simply....... get your-self organized with a grazing chart and a weekly planner.

A grazing plan is not romantic, like going to a bull sale, nor does it seem like you are doing something quantifiable. The grazing plan is the simple counting of days. You don't even have to know much about grass. All that is required is to know the approximate days of recovery for your area, counting the number of paddocks you have (I recommend at least thirteen), then counting the number of days of recovery for a particular time of year.

It is simple to develop a grazing plan, and it is imperative for long term pasture improvement. This must be done before you learn about things like bugs in the soil, best growing grass, or best grass finishing genetics. If you don't have your grass management figured out, everything else you do will be a waste of time.

At that 2000 HRM course, we were taught that grass has three stages of growth. The first stage is slow-growing and highly palatable. The second phase is fast-growing because of the increased leaf area available for photosynthesis, and it is also palatable. The third phase is slow-growing because the plant is putting its energy into reproduction, and it is much less palatable.

For the first four to five years, I grazed in early Stage Two, but I noticed the livestock didn't get superior gains until August, when the grass hardened. I moved my grazing to late Stage Two, where gains improved and we still had good recovery of the grass.

The goal of the grazier, then, is to keep grass in late Stage Two. By employing this simple concept, I achieved significant results. Within two years, carrying capacity doubled. This phenomenon repeated on each piece of land we rented. At one point, we had over five thousand acres of rented land and three thousand head of cattle, and the results were the same on each parcel.

There are subtleties of grass management that become apparent when you consistently use a grazing chart. For example, recovery time varies depending on how much sunlight is available as the growing season progresses. Daylight hours at the latitude of Athabasca, Alberta, (54.7N) are around 4:30am to 11:00pm in June. This means that plants have a lot of time to collect solar energy and grow very fast. To clip them before they mature, you must move your stock through the paddocks rapidly. Another subtlety: if a plant reaches maturity before being clipped, it is my observation that you lose around 50 percent of the possible regrowth. Once a plant reaches maturity, it has completed its life cycle for the year and there is no longer an urgency to grow and reproduce. Again, a grazier's goal is to keep the grass sward in late Stage Two. (I will explain this more in the chapter *Grazing In The Sweet Spot*.)

Everything I have shared so far leads back to the importance of the grazing chart. Your plan is a visual reminder of recovery time. Before starting your grazing plan, though, number all your paddocks. I find numbers are better than names because it tends to be easier for summer students or new employees to figure out which paddocks are which. When I subdivide paddocks with permanent fence, I like to identify each division with a letter (ie. 4a, 4b, 4c, etc.).

At the latitude of Athabasca, thirty-five days of recovery between May 15 and July 15 is pretty consistent (plus or minus five days). After July 15, recovery is more like forty to fifty days. Once you know the Recovery Period for your area, you don't need detailed scientific knowledge about grass to be a successful grazier. Just count the number of days between when you last left a paddock and when you go back to that paddock.

In the case where your Recovery Period is thirty-five days, if you are between thirty and forty days before going back to a paddock you are within the acceptable range. However, if you notice your Recovery Period is more than forty days, you need to rework your plan because you will lose around 50 percent of grass production on the next rotation. If your Recovery Period is less than thirty days, once again rework the plan because you will be injuring the plants.

The number of days in a paddock, or the *Graze Period*, is very important as well. During slow growth, your graze period will lengthen in relation to how much your recovery period increases. Conversely, your graze period decreases when the recovery period decreases. At the latitude of Athabasca, Alberta, during the first rotation in the spring, plants will grow enough that animals can take a bite of new growth on day four following a grazing event. Therefore animals should not be in a paddock longer than three days. This may not seem like a big deal, but the plants are being injured and thus weakened. If we were injuring our livestock, we would certainly notice right away, but plants are easily overlooked. I consider plant injury the same as injuring livestock: *don't do it!*

Here is an example of how to rework a grazing plan. Let's assume it is June 5, and it has been a dry spring. Your grazing plan is based on thirty-five days of recovery, but you notice the paddocks that have already been grazed are not recovering as fast as you anticipated. The recovery period must be increased, so you decide to extend it to at least forty-two days.

There are two ways to do this. The first option is to add more paddocks, which may require some creative thinking (e.g. graze bush paddocks, hay land, road allowance, etc.). The second option is to increase your graze period, which is possible if you have been leaving an abundance of grass in each paddock. If you increase your graze time by one day on ten paddocks, you get an extra nine days of recovery. Of course, if there is not enough grass to increase your graze period then method one is your best option.

Once you decide how to increase recovery time, go back to your grazing chart and erase your plan. Fill in your new plan using the extra paddocks or increased graze periods. Once the chart is complete, count the number of days between when the herd leaves the current paddock and when the herd will be back for the next graze. If the number of days is between thirty-eight and forty-six days, you are set. If it is more than forty-six days or less than thirty-eight days, you must do some more tweaking. Note that you want to avoid too much rest as much as not resting enough.

A Grazing Chart is really just a tool used to get yourself organized. In my days before ranching, I worked on oil well drilling rigs, where I learned to use another valuable tool to improve my organizational skills. What amazing tool could that be you ask? ***Finish one task before starting the next one.*** I learned this skill while working as a motorman (the head roughneck).

When I first got set up as a motorman I was working my butt off the entire twelve-hour shift. The only problem was, I wasn't getting anything accomplished!

The rig manager, Duane Carol, watched me for the first week without saying much. He then called me into his office and asked how it was going. I told him it was a lot of work, and I didn't feel like I was getting anything accomplished. He readily agreed! He explained that to get ahead in my duties, I needed to finish one task before going on to the next. By following that philosophy, I wouldn't have to return to tasks later, nor would I have to fix something right before I used it—it would be completely operational ahead of time. Amazingly, my job became easier and easier as I employed this principle.

Since my rig days, I have continued working on my organizational skills. In that pursuit, I am now using a formal weekly and monthly plan to accomplish my goals. They have helped me immensely while ranching.

Since there are so many variables when ranching, you may question the efficacy of this thinking. You may also feel there are just too many fires to put out to effectively plan. This may be true when you first start. However, the more you use a weekly and monthly plan, the fewer emergencies you will have to deal with, and the more you will be able to get ahead of upcoming duties. Combining the grazing plan with a weekly plan enables you to remember well in advance that a fence requires fixing or a water line must be set up.

When I first started grazing, I was very impressed by all the grass I could grow. I thought that what I was doing was pretty amazing. After about eight years, though, I realized all I was doing was being organized. It was a humbling epiphany. I started looking for other ways to improve worn-out pastures, and I will discuss those methods in the Pasture Rejuvenation chapter. However, had I not been using a grazing chart and got a handle on my grazing, I would not have been able to get to that realization. Therefore, the first thing to do when grazing is get your-self organized, and the first step to being organized is using a grazing chart for planning.

A grazing chart is a simple thing. Kids in primary school can figure it out once they understand the three phases of grass growth; all that is required is to count the days of recovery and the days of grazing. I believe that a grazing plan is the most effective way to improve your pastures. Only after you have that mastered is it time to look at other things like improved seed varieties, the latest chemicals, or any other pasture amendment. Quite possibly, by then you will realize you don't need much more than a grazing plan to improve your pastures.

THURS	FRI	SAT	SUN	MON	TUES	WED	THURS

CHAPTER 5

Determine Your Recovery Period

THE MAGIC OF THIRTEEN

There are two grazing principles that must be adhered to if you are to become a successful grazier. The first is the Graze Period, and the second is Recovery Period. Before I continue, let me jog your memory; Recovery Period is the time it takes a plant to re-grow once it has been grazed and Graze Period is the time animals are left in a paddock. When grazing, neither of these principles should be violated. How do you determine your Graze Period and Recovery Period, though? Let me share with you my method for answering these two questions.

When learning how to graze, I was taught in my Holistic Management (HM) course that you first had to know your recovery period. Once you decided on the recovery period, you then counted the number of paddocks you have, subtracted one paddock because the animals are grazing in one paddock, then divided the recovery period by the number of paddocks resting. Here is an example:

If you have 15 paddocks, that means 14 paddocks are resting (remember the animals are grazing in one paddock). If your Recovery Period is 40 days, you then divide 40 days by 14 paddocks, which gives you an average of 2.8 days in each paddock. Since 0.8 days is a bit difficult to deal with, let's round that up to three days per paddock. Using that information, you can then go to your grazing chart and plan out the moves around the ranch.

You may notice that using this method is pretty easy math and pretty easy to complete a grazing chart. I still use this method after more than twenty years of filling out grazing charts. The problem with this method, though, is how does a person determine an adequate Recovery Period?

If your Recovery Period is too long, you will end up grazing Stage Three grass. If your Recovery Period is too short, you will be grazing grass that hasn't had a chance to replenish its roots and leaf volume. Certainly you could ask an expert grazier or Ag Extension agent in your area, but they will only give you an average for the area, not the Recovery Period for your land. Further, do they have the same grazing goals as you?

I have mentioned more than once that my grazing goal each year is to stock pile high quality, high-volume grass for late fall/ winter grazing. To do this, I need to keep grass in a vegetative state the whole growing season. You may also recall, I call this the *Sweet Spot*. The *Sweet Spot* is when 15-20% of plants in a paddock have gone to Stage Three and the rest are still in Stage

Two. Basically, I want most of the forage to be in late Stage Two growth.

To determine a Recovery Period that allows me to keep grass in the Sweet Spot, I have discovered, through serendipity, that using the Graze Period and thirteen paddocks gives me the correct Recovery Period. For some reason, thirteen paddocks is the magic number. I don't know how it works out that way and I don't know if there is a mathematical theorem to support it, I just know it works.

Somehow, using Graze Period and thirteen paddocks gives me the correct Recovery Period to keep grass in the Sweet Spot if I then follow the HM method of planning. I know there will be a lot of sceptics to this idea and unfortunately I am unable to defend my supposition with any plausible arguments. The only argument I have that it works is my experience. The key though, is you must first determine the Graze Period for your land under your current heat, moisture, and sunlight conditions.

To find out what my Graze Period is at a new property, I first select a paddock that I can strip graze. I look for a paddock that I can easily run a poly-wire from one permanent fence to another and also has enough forage to be grazed for 5-7 days. If you don't have a paddock with enough forage for that many days, just use a smaller group of animals for the few days of the experiment. A critical aspect of my method is that the paddock must have access to water at only one end.

Once I have picked out the ideal paddock, I set up the first break (strip) with enough forage to graze the animals for one day. You are going to have to estimate the amount of forage required, but don't get stressed. Just do your best because if you are off you can always adjust tomorrow. I then let the animals into the first break and let them graze. The second day, I do the same thing. There is no back fence because the animals have to walk through the first break to get to water. On the third day, I once again do the same thing. Here, though, the animals have to walk through both the second and the first break to get to water.

I follow this pattern until the cattle stop in the first break to graze. In our area, during May and June, the animals stop in the first break on day four, which means my Graze Period is three days. The Graze Period is three days because there is enough new growth by day four that the animals can take a bite of new growth. You probably already understand this is a BIG NO-NO!

Your Graze Period will change as conditions change. As an example, in 2021 our spring started out normal, so I used a normal Graze Period for that time of year, which is three days. By mid-June we were very dry and unusually hot. When I did the strip graze test again, the animals didn't stop in the first break until day six. That meant my graze period was five days.

I share this example to make the point that your Graze Period will change as conditions change. Interestingly, that fact is rarely acknowledged. What is widely discussed, though, is that Recovery Period changes as conditions change. However, to be a successful grazier, both Recovery Period and Graze Period must be respected. The system doesn't work if one or the other of these principles is violated.

Once I have my Graze Period identified, I can then determine the Recovery Period using the HM method I discussed earlier. To do this, I assume each herd has thirteen paddocks to graze. Therefore, if the graze period is three days, then the Recovery Period is thirty six days. Let me demonstrate:

- *13 pad's – 1 pad = 12 pad's (remember the animals are in one paddock, which leaves 12 paddocks resting).*
- *12 pad's X 3 days = 36 days of recovery**

*This is virtually the same number of days it took me several years to discover through trial and error. Who knew?

Now that the Recovery Period is established, I can employ the HM method to determine the number of days in each paddock when I'm dealing with more than thirteen. To explain this better, let me share an example where a herd has twenty paddocks to graze, remember though, Graze Period is a maximum time frame not a minimum:

20 pad's – 1 pad = 19 pad's resting

Assume:
Recovery Period: 36 days
Graze Period: 3 days

- 36 days divided by 19 pad's = 1.9 days (round up to 2 days)

In this example, the average time in each paddock will be two days. That does not mean each paddock must be grazed for two days. In fact, some paddocks may be grazed for one day, while others may be grazed for three days. The important point is that no animals go back into the first paddock until 36 days has elapsed and no paddock had animals grazing for longer than three days. Here, the principles of Recovery Period and Graze Period have not been violated.

I have spent a lot of time trying and testing out various grazing methods. Discovering that thirteen paddocks helps determine an accurate Recovery Period was a fluke. Through trial and error, I had already established the Graze Period and Recovery Period for my area. The lucky part came when I took over a pasture that had thirteen paddocks. Half-way through that growing season, I realized the numbers worked out evenly on my grazing chart. I was surprised by how simple it was to plan out my grazing rotation.

Since that discovery, I have tested out my theory several times with the same positive results. In hindsight, I wish I would have known about the Magic of Thirteen when I first started. Thirteen paddocks is where I could have started to learn how to graze. Instead, I was concerned with plant species, how much forage each paddock produced, soil types, bugs in the cow pats, etc.; it was overwhelming! Yes, those things are important and interesting, but those discoveries could have come after I learned how to keep my grass in the Sweet Spot.

I know this may sound blasphemous, but I needed to be economically viable and as a custom-grazer, my income was tied to the amount of grass I produced. It may have been different if we had been at a later stage of life and we didn't have young kids to raise and a mortgage to service.

Learning about soil biology, soil types, plant species, etc is certainly interesting and I continue to learn more about that stuff with each passing year, I'm just saying that at the start, what I really needed to know was how to grow more grass. Using The Magic of Thirteen is a simple and practical way to improve grass production. It enables me to quickly assess a ranch's grass potential, and I have used it on each property I manage.

Several people have already challenged me on this idea, and I am certain there will be many more, so let me demonstrate why other paddock numbers don't work as well.

In the following table, I use a variety of Paddock numbers. If we assume that the Graze Period is 3 days and the Recovery Period is 36 days, and since it is my goal not to violate either principle, the only number of paddocks that works is thirteen. (Remember that the principles of Graze Period and Recovery Period must not be violated. Let's assume 36 days Recovery Period and 3 days Graze Period. Please remember these numbers are just for example sake and do not represent Graze Period or Recovery Period for your area.)

To explain further, if we start with the recovery period (36 days), then 8-10 paddocks violates the graze period; 12 pad's is close, but does not provide much flexibility, and 13 pad's is just right.

If we start with the graze period (3 days), then 8-10 pad's violates the recovery period as does 15-20 pad's. Thirteen is the only number of paddocks that lets you adhere to both the graze period and the rest period. Thirteen paddocks should really be called the Goldilock's Number not the Magic Number... not too many, not too few, just right!

The Magic of 13 Paddocks

# Pad's	Graze Period	Recovery Period
8	3 days	21 days
10	3 days	27 days
12	3 days	33 days
13	3 days	36 days
14	3 days	39 days
15	3 days	42 days
20	3 days	57 days
8	5 days	36 days
10	4 days	36 days
12	3.2 days	36 days
13	3 days	36 days
14	2.8 days	36 days
15	2.6 days	36 days
20	1.9 days	36 days

More than fifteen paddocks

Greater than fifteen paddocks may sound great, but it has been my experience that when I have more than fifteen pad's, it is very easy to get too long of a Recovery Period. Further, to maintain integrity of the two grazing principles, it requires unnecessary labour. Why move animals every day when once every three days accomplishes the same result?

You may have noticed that thirteen paddocks is the minimum number required to respect the two grazing principles. However, more than thirteen is perfectly fine as long as you use a grazing chart to keep yourself on target of the Recovery Period. Personally, I like fifteen paddocks because I have more flexibility for managing paddocks of different forage production.

Often I will encounter paddocks that have an abundance of forage, while others on the same ranch produce very little. If there are more than thirteen paddocks, I don't have to change anything because the animals can stay the maximum Graze Period in paddocks with abundant forage and only a day or even half a day in ones with little forage. As long as I use my grazing chart, I can stay on track for the proper Recovery Period and still get all the paddocks clipped so they will stay in the Sweet Spot.

A cool thing about using The Magic of Thirteen to figure out your Recovery Period and Graze Period is that it gives you information specific to your place. To formulate your initial grazing plan, you may need to use recommendations from people in your area. However, once you are going, this tool can help you check those recommendations against what is actually happening on your place.

I mentioned earlier that this technique only works if you want to keep your grass in the Sweet Spot. If your goal is to let your paddocks go into Stage Three, then it will not be useful. For me, though, my goal each spring is to grow high quality, high-volume grass for late fall/winter grazing. The Magic of Thirteen has helped me achieve this goal by simplifying what took me many years to learn through trial, error, and observation.

By now, you know I want things to be simple. Thirteen paddocks is simple, and it doesn't matter the paddock size, what forage is growing, or anything else for that matter. All that is required is you know how to do simple math and you own a grazing chart. Once you have those two items handled, get grazing. The rest will follow.

Infrastructure

CHAPTER 6

Water

Providing plenty of clean drinking water for livestock is as critical to a grazing system as plentiful grass. Too many times I have cheaped out in some area of a water system, only to be in crisis mode later in the season. I have messed up with poor water sites, by relying only on snow, and with flimsy water troughs. Fortunately, I eventually learned my lesson: build a robust system the first time!

There are many ways to provide water and any agricultural extension agency will have all the information you require. However, I want a system that is simple, easy, and portable. I developed the one I will share with you after seeing and using numerous water system components. If they didn't fit my criteria, they were gone.

As an example, solar waterers are still being promoted as a fantastic delivery system. I used solar systems over fifteen years ago and they were finicky and unreliable. They are still finicky and unreliable! I know people will bash me over this point; however, if a twelve-year-old can't fix a water problem, I don't want that component in the system.

In my experience, pipelines are the most cost-effective water distribution component. Picking on solar again, for the cost of a solar system capable of watering 150 cow/calf pairs, I can buy 1.5 miles of two-inch DR17 pipe. An added benefit is that the person operating the pipeline only has to know how to open a valve! The same can be said of building a dugout. That money can buy quite a bit of pipe.

I have broken this section on water into three parts: pipeline, trough, and water site. In each category, I will explain my "best solution." These will not be the only option by a long shot. However, I have tested each component through trial and error with the underlying principle that it must be simple, easy, portable, and robust.

If a component can fulfill these requirements, it will also enable a person to expand their operation without another outlay of cash. As an example, if you start with only a one-inch water line and 5GPM flow because you have 80 cows, you will always have difficulty if you expand to 300 cows. Most people tell me they only want to run 80 cows, 120 cows max, which is why they only require one-inch pipe. That is fine, but in the next breath they tell me they are getting geared up to stay at home and not work off farm. In my head, I'm thinking, you'd better figure out how to run a lot more than 80 cows to pay your bills!

PIPELINES

The goal for any rancher should be good, clean stock water, and ample quantity at high volume. For me, twenty gallons per minute is my target rate of flow, allowing me to water a lot of livestock at once. When you have 20 gpm flow to start with, you won't be scrambling to solve water issues when your herd expands or there is a failure in the system.

I also strive for 20 gpm because it allows my system to be portable. If you have a high-flow system, your trough does not have to have a large capacity, which makes it light enough to move with a four-wheeler or a horse. If the flow is low, then when a group of animals comes to drink, there will be a lot of fighting. This will put a lot of strain on your trough, so it must be very robust.

As I mentioned earlier, I'm a fan of pipelines; more precisely, two inch DR17 pipeline. DR17 is more robust than regular water line and surprisingly less expensive. However, normal hose clamps tend not to be adequate to hold joins together. The connectors and fittings I use are a compression fitting called Cepex. The more I use these fittings, the better I like them. They have the same integrity as a butt fusion and also have the flexibility to undo the fitting if you want to make a change to your system. Of course, you pay dearly for the flexibility, but I assure you it is worth the premium.

When designing a pipeline, I first figure out how to get a flow of 20 gpm. If your well does not have that capability, then you could look at a dugout or a large storage tank. If your well has lower flow, but has enough to water your livestock over a twenty-four-hour period, then you require a storage tank to hold enough that the stock are not relying on the well when watering. A dugout works the same way, but it must have enough capacity to get your stock through a droughty summer.

Next, consider how to charge your pipeline. If your source is on a high elevation relative to the rest of the property, then gravity can work very well. You must check the elevation with Google Earth first though to make sure you will have enough hydrostatic pressure to overcome the friction loss in the pipe and provide enough flow to the trough.

For quick calculation, one metre of elevation change is approximately 1.4 pounds per square inch of pressure. (Don't hate me because I use both metric and imperial measures! It's how I was raised.) I would recommend getting help from an expert agricultural extension person to figure out friction loss, because it can be a big issue on long distances. Once you know the location of your water source, it is time to hit Google Earth Pro.

Google Earth is a great tool because you can play around with different routes before laying any pipe, and it shows contours that may help or hinder you. Plan your route so that the pipe runs along fence lines. This will protect the pipe from being damaged by equipment. Equipment running over pipe in the winter has been the only thing to damage my aboveground pipeline. Although, a beaver once chewed three holes in a 1¼-inch line we had crossing a small pond, and I have heard of coyotes chewing holes in one-inch line. FYI, coyotes are very hard on plastic handles of PVC valves. I don't know why; they just like to chew coloured plastic!

Once you have your plan, show it to a few people for their input. You might be surprised by what they see.

Before we leave this topic, I had better mention that there is no need to bury your pipe unless you are creating a winter water site. I designed my first pasture pipeline in 2001 and that pipe is still functioning today without damage to the line. Valves have broken when someone drove over them with a piece of equipment, but the pipe itself has not failed even when the line didn't get drained early enough in the fall and froze completely. By not burying the pipe, you have greater flexibility for making changes. Also, in northern climates, the pipe will be ready to use much earlier in the spring.

As for pumps, my preference is a variable speed pump. Again, I would recommend getting some help on sizing the right pump. If you don't have elevation to deal with, a pump that puts out 20 gpm at 40 psi should be more than adequate. Variable speed pumps generally cost more money, but they are well worth it. They save on electricity and I like how they ramp up as the demand increases.

Incredibly Awesome Anecdote

Before Google Earth, I used a handheld barometer to calculate elevation change. In 2004, when designing a gravity flow system for my neighbour, I calculated that the pressure at the last valve—which was three miles away—was only going to be 6 psi. When a GPS salesman showed my neighbour how to use GPS, he calculated that there would not be enough elevation change to provide the 6 psi I had calculated. The pipe was all laid out with valves installed ready for the next spring. I fretted all winter that my calculations might be off, even though I knew GPS elevation measurements were not that accurate. Thankfully, my barometer did not lie, and the system worked just as designed!

GPS accuracy has improved immensely since then, which is why I now use Google Earth.

STOCK TROUGH

Once you get to herds over two hundred and fifty animals, particularly yearlings, they seem to find every weakness in your delivery system. It has taken me twelve years and numerous prototypes to come up with a trough and valve that is yearling proof (knock on wood). This trough is light enough to be moved with a horse or four-wheeler, can be accessed by calves and sheep, and is robust enough to withstand cattle fighting to gain access. My favourite valve is a one inch Jobe valve, because it allows high flow and has been more reliable than any other valve I have tried.

To get water from the pipeline to the trough, install one-inch PVC valves at a spot under a fence. Install a camlock with locking ears into the valve. From there, use a one-inch polyethylene pipe thirty feet long to go to the trough. String the one-inch line under the skid of the trough starting at the end furthest from the outlet, which is located under the trough. This allows the line to have some movement without breaking the fitting (a ninety-degree elbow and camlock). I have tried a number of different materials to connect to the trough, and of the ones tried, one-inch poly has worked the best to avoid being pinched off, getting twisted, or getting punctured.

WATER SITES

Water sites can be a tricky matter. For groups above three hundred head, a site thirty paces by thirty paces seems to work well. There is enough room for a group to come drink together, yet not enough room to loaf. To maintain a solid base in the water site, wood chips perform very well. The pad must be at least twenty-four inches thick, though, and be big enough around the trough to handle three animals deep. The cool thing about chips is that a herd of eight hundred or nine hundred yearlings can water without pounding out the pad. The pad also slopes gently into the surrounding ground, so there is no shelf created at the edge of the pad. When it rains or the trough overflows, the pad stays dry and the water soaks through the chips out to the sides of the pad.

To get to the water site, an alley is my preferred choice. I did not always think this way. Originally, I was taught that alleys create a lot of damaged ground, so we moved the trough around from paddock to paddock. When animals lift their heads from drinking, water drips from their muzzles. Some animals also purposely splash water out of the trough. Both scenarios create a mud hole that gets tramped deeper by the herd. Once we started increasing herd size, damaged spots would show up after each trough move.

An example pad, although it is a bit small. We had to leave it as is because it started raining and never quit all summer. We couldn't get a truck in to add more wood chips!

To alleviate multiple spots, we decided to only use one site, positioned to provide access from several paddocks. That progressed into using alleys. As an added benefit, we noticed that labour really decreased with alleys because all we had to do was open and close gates The troughs didn't have to be moved very often, and there was much less infrastructure to maintain.

For alleys to flow really well, they should be seven paces wide. Our alleys started out only five paces wide. After building a system for a neighbour, he complained because it was difficult to drive a pickup down the alley. Since then, our alleys became seven paces wide to provide a bit more flexibility.

One last thing about water: draw out a map of your system. The map should contain valve locations as well as the route of the pipeline. This is very important, because once grass grows over the line it becomes difficult to find. A map will save you a lot of cussing and swearing!

CHAPTER 7

Fencing

Fencing for me has evolved a great deal since 2000. Our first fencing used rebar posts because they were half the price of step-in posts. We were building and taking down two acre paddocks every two days, dragging the rebar around in a modified manual golf cart. We were in great shape, but we were exhausted! We had forgotten to factor our labour into the cost analysis.

Labour costs more than just the time it takes to perform a task. You can easily track the hourly rate. The part that is hard to calculate is the mental and emotional cost of unnecessary labour. Ranching should be fun and easy. When we are run off our feet and always exhausted, our thinking becomes impaired. When I am in an extended period of high labour, I get tunnel vision and my creativity is severely impaired. Plus, it's not that much fun!

Once I realized the hidden cost of labour, I became obsessed with reducing it. I want everything to be simple and easy, and the best way to accomplish that feat is to once again ask the question, "Can a twelve-year-old do this?"

Initially, our permanent fences were one strand of high tensile, and it worked well until we expanded the custom grazing operation. Once we started taking in yearlings, there would be times when one or two head would jump the wire, or duck under into another paddock. It would take twenty to thirty minutes to gather those yearlings into the proper paddock.

Thirty minutes is not much time, but it was time that could be better spent doing something else. That's when I decided to start running two strands of high tensile. Afterward, we rarely had animals in the wrong paddock. You may not think this is a big deal, but let's calculate how much labour the above scenario takes up in a grazing season.

When we were custom grazing, we ran between 2500 and 3000 yearlings in five to six groups. Usually the cattle started arriving May 15 and would be gone September 30. That is twenty weeks. If we had to gather cattle from the wrong paddock three times per week; that is sixty instances of gathering cattle.

I already pointed out that it usually takes about half an hour each time, so sixty incidents times thirty minutes is thirty hours of wasted labour. Multiply that by $18 an hour for a summer student, and you have just wasted $540. It may not be much, but often it takes a summer student longer than a half hour, if they are able to gather the animals at all. When they are unable to accomplish the gather, guess who has to go and do the job? It's someone who would rather be playing poker with his retired neighbours!

Here is the template for building electric fence Tom Krawiec style:

Supplies

- - 12 gauge high tensile wire
- - 12 gauge double-insulated wire
- - Kencove electric bungee cord for gates
- - claw insulator
- - Kencove or Stafix single-throw cut-out switch
- - Kiwi or Daisy in-line tightener

Instructions

1) *Use 8' by 5-6" pressure-treated posts as end posts and gates pounded halfway into the ground. Build gates twenty-one feet wide.*

- Explanation: Larger end posts don't break when moose or elk hit the fence at full speed. When they are pounded in halfway, a brace is not required to keep the post from moving. Gates are twenty-one feet (seven paces), which is wide enough for large groups of animals and most equipment to drive through.

2) *Use 7' X 4-5" pressure-treated posts as line posts pounded three feet into the ground, spaced sixty feet apart.*

- Explanation: Posts are sixty feet (20 paces) apart to reduce sag. We started out at ninety feet because that was what we learned in HRM, but we found there were too many issues with animals getting into the next paddock.

3) *Install two high tensile wires, placed at forty-five inches and twenty-eight inches from the ground.*

- Explanation: Install two wires because it will keep jumpers from jumping and sneakers from popping under into the next paddock. The height of the lower wire is important because it allows calves to get under the wire to "creep-graze" and allows newborns to come back to the mom when they get on the wrong side of the fence. Young calves can crawl through a five-strand barbed wire fence but have a heck of a time getting back, so this reduces the hassle of going and fetching young calves.

- If you are running sheep, one more hot wire and a bottom non-charged wire seems to work quite well with hair sheep. If you are running feeder hogs, you just need a third hot-wire.

4) *To tie off at the end post, use three insulators positioned so the wire does not contact the post. It is easiest if only one screw is installed in the side insulators so they can be turned when threading the wire. When making the final tie, do it as close to the post as possible and make fifteen to eighteen wraps. These will hold when a moose or elk hits the wire at full speed.*

- Explanation: I don't use S-strainers because they are relatively expensive and it is another two ties, which takes more labour. I like to have a "hot" wire going around the post because it keeps animals from rubbing against it. If there is a rush for the gate, it slows movement through the gate because animals that contact the post get shocked and back off.

- When tying off, use more wraps than fewer and do not do a Kiwi tie; the Kiwi-tie breaks and fewer wraps unravel when moose and elk hit the wire.

5) *Tie the two wires together using a short piece of high tensile on the side where the flow of power begins. This will make the two wires act as one wire for the flow of electricity.*

- Explanation: Many years ago, the Gallagher rep told me to tie the two wires together, so they act as one wire and allow more electrons to travel down the wire with less resistance.

6) *Place a tensioner on each wire fifteen feet from the end post where the electricity begins.*

- Explanation: Always place tighteners in the same place on the wires. It saves time looking for a tightener when adjusting tension.

7) *When installing a cut-out switch, place the switch on the first post of the "fence leg," not on the main line fence. Use double-insulated wire for the tie-ins.*

- Explanation: Use actual cut-out switches so anyone can turn off a section of fence. For a long time, I just used alligator clips or only had one or two cut-outs. Alligator clips get pulled off by calves or chewed, which leads to inadvertent shocks. Not fun! Before I installed multiple cut-outs, fences would get repaired while there was still power on the fence. The repair was only a temporary fix that became a long-term fix. Not good!

8) *Cut or break off all wire "tails." This will help prevent shorts.*

9) *Use double-insulated wire and L-clamps when joining two sections of fence.*

10) *Gate handles: cut eighteen inches of insulated wire, tie a bow in the middle, strip 1.5 inches at each end, and then bend one end into a hook, the other into a circle.*

WATCH: Building Gate Handles

You can find this and other videos at simplyranching.ca/video.

11) *Gates: Cut bungee cord one-third of the gate width. Use turbo wire for the rest of the gate. Install short pieces of red tuck tape spaced out on the wire.*

- Explanation: I like homemade gate handles because they are cheap and reusable. If an animal goes through the gate, the hook straightens and the handle is not broken. To repair the handle, you just bend the hook again and presto, your gate is perfectly functional!

A few years ago, I started using a short piece of electric bungee (I really like the Stafix bungee) tied to turbo wire for the gates. There tends to be sag if the whole gate is bungee, but with a short piece, you get plenty of flex without the sag.

As a final touch, I like putting several strips of red tuck tape on the gate wire. It seems to keep both domestic and wild animals from going through gates. Further, you can easily see if the gate is open from a distance. I know lots of people swear by poly tape or rope. It has been my experience, though, that the tiny wires break over time and the gate loses its ability to conduct electricity.

In my experience, this type of fence needs very little maintenance because of the sturdy materials used. It may not seem like much, but every broken post takes at least an hour to replace, not including the mental capacity required for the repair.

Before I finish, let me state that gates are good. Gates allow for easy flow of animals and fewer headaches moving animals from paddock to paddock. If there is an issue with animals not flowing freely, cut the fence and build a gate. When braces are not required, building a gate is quick and inexpensive. I like underground gates rather than overhead because invariably, one day someone will want to drive tall equipment through the gate and take out the overhead wire.

This template is not the cheapest to build and is not in everyone's budget. If it's not in your budget, then use it as a target. Really, though, when you calculate labour costs, it is quite economical.

Incredibly Awesome Anecdote

When you have this type of fencing in conjunction with a robust water system, ranching becomes pretty simple. A rancher I know, (she runs around 350 cows, has four children under thirteen years old, and a husband who works off farm), told me how easy it was to run her herd with this type of fencing and an extensive water pipeline system. I think that is pretty cool. Again, ranching should be simple, easy, and fun!

You can find this and other videos at simplyranching.ca/video.

Animal
Handling

CHAPTER 8

What is Low-Stress?

I have been a Bud Williams follower since 2000. Since then, I have also taken a course from Dylan Biggs. From Dylan and Bud, I learned how to stop, turn, and speed up livestock. The animal handling principles and techniques I learned are very similar for different species, and I have been able to hone my skills with cattle, sheep, hogs, horses, and turkeys.

One thing I struggled with was teaching other people these techniques. For some reason, I was unable to explain what I was doing in a way that summer students, owners, family, friends, and apprentices could understand. Luckily, I discovered the writings and videos of Whit Hubbard. He does a fantastic job explaining the basic skills required to become an accomplished low-stress animal handler. Take the time to look up his work.

So what exactly is low-stress handling? Often, it is the idea that animals are always at a walk. Sometimes, it is the idea that if animals come when you call, then that is low-stress handling. Both these statements may be true to a point. However, in my experience, low-stress handling involves much more.

To my thinking, low-stress handling is the ability to start, stop, turn left, turn right, speed up, and slow down an animal or a group of animals. Further, it is the ability to perform these maneuvers in a controlled manner. It is my belief that until you can accomplish these tasks, your animals are not truly under control.

The cool thing about having your animals under real control is that you can go virtually anywhere with them. You no longer have to dread moving livestock. I have been moving animals in small groups, large groups, and individually multiple times a week for more than twenty years and I still get a kick out of seeing how precise I can be with my motions. Once you get to this point, planning your grazing rotation becomes much easier. You can trust your ability to move animals wherever and whenever you want, enabling you to accomplish your grazing goals much easier.

As I pointed out earlier, I think Whit Hibbard does a fantastic job explaining the basic techniques, and I will not try to improve on his stuff. What I will share, though, are a few skills I use to graze effectively.

THE WAVE

Since 2006, I have been working on an efficient method to move newborn calves and lambs from pasture to pasture. It is something that scares many people and keeps them from managing grass at the same time they are calving or lambing. Most people will have a birthing pasture that gets hammered because the animals are parked for three to four weeks. By staying put, the newborns are being exposed to ground that is contaminated with mud, urine, and dung. If there are sick calves in the bunch, newborns are also exposed to those contaminates.

Another problem with not moving is that the grass in other paddocks is growing and becomes mature by the time a person starts the first grazing rotation (if you are ranching like a twelve-year-old, you are calving and lambing on grass!). However, there is a method to move newborns effectively, and I call it "The Wave."

The first step in the Wave is to open the gate to the new paddock. This way, the herd can saunter through the gate instead of bunching up. If they bunch up at the closed gate, there is a universal tendency to rush through once it is open and forget about the calves and lambs.

Once the gate is open, the handler moves to the back of the group and begins a zigzag movement (Whit Hubbard describes this very well in one of his articles). The zigzag is very slow.

The handler should stop frequently and let each mother get up. Once she is up and has gathered her young, it is time to proceed past her on to the next mother. When the handler reaches the edge of the group, they turn around and go back on the initial path. This movement will generate some forward movement, but not a lot. That gentle movement will in turn create just enough energy to get the next "row" of animals up and gathering their young. Continue this pattern until the group is moving. Of course, with any young that are still wet, you can veer off and let them mother up. Those mothers will follow as soon as the newborn can walk.

The Wave takes a lot of patience, and since it takes "as long as it takes," don't be thinking about all the other things you have to do that day. I am an ardent believer that the energy you bring when moving animals is the energy the animals feel. Therefore, when moving newborns, make sure you feel calm and relaxed.

As long as the group is moving, let them move at their own pace. They dictate the pace, not you. If you are on horseback, an old, semi-retired horse is the one to ride for this chore. When moving a distance of over one mile, let the group stop after each mile and rest so the young can catch their breath and suckle for thirty minutes or so.

It has been my experience that animals respond very well to verbal cues. If the group of animals is trained well as a cohesive unit, I will shout "Hup! Hup!" when I want them to move. This trains the group to prepare for moving when I holler, and not be bothered when I ride through checking for birthing trouble. The mothers learn quickly that when they hear me holler, it is time to move, but not when I am just riding through. After the young are about six weeks old, I will start using a whistle to call the group into the next paddock. This training helps immensely when grazing in bush paddocks.

One final note: gates should be positioned such that the animals can go straight across from one paddock to the next. This prevents young animals from getting trapped walking down the wrong side of a fence because they can see animals on the other side. It is very important when crossing a road or swamp and will make your life much easier. I know it is a pain to build another gate, but trust me. Just do it!

WATCH:
The Wave

You can find this and other videos at simplyranching.ca/video.

CHAPTER 9

Creating a Herd

One of the skills I have developed is the ability to create a herd (when I use the term "herd," I also mean a flerd, MOB, or flock). When I first began grazing, I thought animals were a herd when they were grazing in small paddocks. Just because animals are in proximity to each other does not mean they are a herd; it just means they are grazing the same grass. To create a herd, you must teach animals that they prefer to be with their buddies. You create an environment where being alone is unnatural. I will do my best to explain how the heck I do this.

As I stated earlier, I aim to train each group of animals I handle to drive, turn, stop, and work as a unit. Whit Hubbard explains very well how to drive, turn, and stop, and he also explains how to teach individuals to stay with the group. It is on this last skill that we differ a bit in our approach.

Whit has developed a method in which you become an annoyance to the animal that leaves the herd. I, on the other hand, am very aggressive when training "bunch quitters." In my mind, I become a wolf, whether I am on foot, riding a horse, driving a pickup, or driving a 4-wheeler! So you should probably stay out of my way!

This notion of becoming a wolf comes from my observations of a wild herd of bison in the Northwest Territories, Canada, and videos of barren lands caribou and wildebeest. Caribou, wildebeest, and wild bison are trained by predators that their survival depends on staying with the group. The slow learners are picked off pretty quickly.

The bison herd I observed knew the importance of staying with the group because there were a number of wolves in the area. As the herd moved past our remote work site, each individual stayed in close proximity to the group. The young bulls were fighting and cavorting about, but always moving down the cut line within twenty feet of the group. At this time, I also saw natural herd effect in action for the first time.

In March 2000, I was working on a crew hydro-testing a new pipeline. Once we had our equipment set up, we went to camp for supper. When we came back, the herd of bison had moved into our work site and "occupied" the joint. We could not take a step without stepping on a bison track. There were only about fifty animals in that herd, but talk about herd effect!

This episode greatly influenced my thinking about how animal handling can affect grass management. Many people think bison naturally herd. However, the tame herds I have seen do not behave in the same manner as that wild herd in the Northwest Territories. This further confirmed my belief that herding-up is something that needs to be learned.

So now you may be asking yourself, "How do I become a wolf?" The following is how I do it.

For this example, let's assume I am riding a horse. I believe a fit, well-trained horse is most effective for this endeavour.

With a new group of animals, I first assess the flight zone. Depending on the previous history, some groups will have a large flight zone while others will barely budge from your pressure. With a large flight zone, you might need to stay one hundred feet or more to gather them. With a small flight zone, you may need to almost ram them to create some momentum. I will not deal with the specific mechanics of how to gather because Whit does a very good job. Just be cognizant that flight zones differ from animal to animal.

When gathering a new group of animals, it has been my experience that there are always bunch quitters. These individuals must be pursued by the "wolf."

To become a wolf, I create a rage within myself. I am angry, tense, loud, fast, and probably pretty scary to be around. I pursue each bunch quitter with this high level of energy and direct them back to the herd. Once they get to the edge of the herd, I immediately stop my horse & drop my energy. Read that last sentence again, because it is *critical to your success*. Dropping your energy is an immediate reward. If you or your horse will not do a hard stop and immediately settle, training bunch quitters becomes much more difficult.

The process of increasing your energy and then dropping it back down is not a natural skill. It takes intention and practice. Like any skill, though, with practice comes proficiency. After a couple of years, regulating your energy will become natural. If you stay consistent with the approach I just described, you will have a working herd within two to three weeks, assuming you are moving your animals two to three times per week. If you stay aggressive and consistent, you will have a hard-bonded herd within a couple months.

Being consistent is very important. I know it can be a pain in the butt when you have plans and a cow decides there is better grass in the opposite direction. If you get after her right that moment, there will be fewer and fewer instances of that happening.

There are a number of benefits to having a well-bonded herd. The first benefit is that moving becomes very easy. Many times I have moved over six hundred animals myself or with just one friend. I vividly remember my daughter at twelve years of age helping me move 950 heifers three and a half miles down a county road to a new paddock. The herd plodded along four by four and was strung out for three quarters of a mile. She was so sweet and helpful back then. Sigh!

You can find this and other videos at simplyranching.ca/video.

There are other benefits that don't present themselves in such an obvious way. First of all, when animals become a herd, they affect the soil with their impact and you don't have to do any fencing to create that impact. Like the Northwest Territories bison at our work site, when animals stay close together, every patch of ground gets covered, without any additional fencing.

Another benefit is that animals figure out how to work together. A few winters ago, I noticed a herd of dry cows walking around a paddock in a tight group. It was very odd, so I went to investigate. The herd was walking together to break the crusted snow into small chunks so they could get to the stockpiled grass. It was amazing to witness. I have no idea how those cows figured out how to work together!

When animals understand they are part of a group, their behaviour changes, and I'm not sure of everything that will happen. I have seen cows walk from a waterer, through a paddock full of bale grazing bales, back to the herd in another paddock of bales. If it was one cow, this would not be of note. However, it was a steady stream of cows going to water, then back to the herd over the course of three hours.

Another time I left a gate open by accident as I scrambled to sort out a water failure. About one hundred pair out of four hundred travelled through four paddocks to a dugout, drank, and then came back to the herd. Those animals walked through "ice cream" quality grass to get back to the herd. I just find this behaviour amazing and the more I witness, the more I am convinced of the efficacy of what I am doing.

The final benefit I notice is also the coolest. When your animals are truly living as a herd, they birth within the group. There are no animals wandering off to give birth on their own. It is just like what you would see watching caribou calving beside the Beaufort Sea.

I have witnessed an anomaly a couple times though, where a group of six to eight cows walked away from the herd to have their calves together. Each time, there was one matriarch standing watch while the others all calved within a couple of hours. Very odd, very interesting, and I have no idea why. If you ranch in a high predator area, you can certainly appreciate how many calves would be saved if your cows calved in a group.

Training animals to act as a group does take dedication and consistency. And yes, the benefits are certainly worth the effort. I realize the "Wolf" method is not low stress. However, this method is only used during the training phase. Once you truly have a herd, handling becomes a joy and a wonder!

CHAPTER 10

Multi-Species Grazing

I t has been my experience that different species can be trained to act as a group in the same manner I described in the last chapter. In fact, multi-species can bond with each other and behave as a single MOB.

My idea of multi-species grazing is a MOB of different animals together at the same time. It is not a leader-follower system. The leader-follower system is similar to traditional crop rotation. Yes, you may have multiple crops, but each one is grown as a monoculture. The big push now is for grain farmers to grow poly cultures. Proponents try to get significant diversity in their crops. In my mind, the same should be true with our animals.

When I began putting different animals together, it wasn't because I wanted to improve the grass, it was simply to save labour and infrastructure costs. The improved grass was just a bonus!

A MOB of cattle, sheep, hogs, and horses.

SADDLE HORSES ARE JERKS

First off, let me assure you I do not hate horses. In fact, this chapter is not really about horses at all; it is about soil and grass and how different species work together. Horses can really improve grass and soil even though the way they are currently managed does the opposite. It's just that when horses are grazed in a group (MOB), they have some peculiarities, as does each species I have dealt with.

There appears to be a hierarchy of species when they are combined. Saddle horses tend to have an air of superiority above all others. It's not even all horses, just saddle horses. Being trained must change a horse's perception of itself, because I have not seen this behaviour when grazing bucking horses.

Cattle, on the other hand, are pretty easy-going and are content with good grass and clean water. They do not, however, appreciate pushy sheep. Yes, sheep are pushy! They butt to the front of the line and don't observe 'proper' social etiquette. For some reason, sheep don't take the hint when a cow bunts them out of the way. Hogs can be that way as well, but are actually the social butterflies of the MOB. They don't really care whom they hang out with as long as there is good grass to eat.

Note: turkeys have not fared well grazing with larger animals. The economic loss from irate cows and curious hogs made me quickly abandon adding them to the MOB! Maybe someone else has been successful doing this, but it hasn't been me.

When putting a multi-species MOB together, you will see these behaviours amplified in the first two weeks. The saddle horses will do things like block a gate after the MOB has been nicely flowing for a quarter mile. As a herder, you will be in the back wondering why the whole MOB has stopped. You glance up to see your favourite saddle horse turned sideways in the gate, blocking any animal from passing. It becomes your job to ride to the gate and chase that horse through with a stick, rock, or something hard to get your point across. Like I said, horses are jerks! Although maybe I am being a bit too hard on horses, because I have had a couple bulls do the very same thing.

For the MOB to work effectively, it requires a strong leader. *You* must become that leader. If there are bunch of quitters, you must harass them and teach them the MOB is their safe place. It doesn't matter what species or what age. Each animal must know they are safe as long as they stay within the MOB. If an animal is blocking a gate, you as the leader, must let that animal know it is unacceptable behaviour. If a cow keeps bunting sheep away from the water, you must get after that cow. It may be difficult if the horse blocking the gate is a pet to you. Out in the MOB there are no pets, only equal members of the MOB. So suck it up and be a strong leader!

After the initial training phase is over, you will see amazing things happen within the MOB. The animals learn they are part of one unit; it doesn't matter what species they are. You can still sort off the animals you want, though. However, if you happen to bring in only the cattle or only the sheep, you may turn around to find that the rest of the MOB is following. Even the horses!

Gathering 250 ewe/lamb pairs out of 300 cow/calf pairs.

THE MULTI-SPECIES MOB IN PRACTICE

Before I proceed, let me share a few cool moments I have witnessed while grazing multiple species as a MOB:

I was bringing in a MOB of over three hundred sheep, thirty hogs, twenty horses, and 180 cattle (steers, dry cows, and bulls) and the group was spread out about a quarter mile down a causeway. At the end of the causeway was a steel gate leading into another pasture. There was a water trough just inside the gate. Several horses had to wait at the gate because cattle stopped to drink. The sheep kept going through the gate into the pasture. At one point, a horse picked up its hind leg to let some sheep pass under. Maybe horses aren't always jerks!

- As I was sorting off a bull on foot, the bull carefully walked through the ruminating MOB without stepping on a single calf, lamb, or hog.

- Many afternoons, I've ridden out to the MOB to find the whole group lounging around together, ruminating. Each time I am filled with awe because, to me, it is the epitome of "Peace in the Valley."

- When coyotes are present, sheep will move to the centre of the MOB as evening approaches. It then becomes very difficult to sort off some butcher lambs if you happen to forget until after supper.

- Once the animals become an actual MOB, the ewes and cows will birth within the MOB and not off in a corner of the paddock or in the trees. Personally, I do not have horses in the MOB until after the first cycle of calving or lambing. Once again, horses can be jerks and will sometimes go stampeding through the group to the detriment of the newborns.

I know the concept of different species bonding may seem far-fetched, however, I assure you it is real! The benefits are huge in terms of reduced labour, improved soil health, and decreased predation. Give it a whirl.

MOB Infrastructure

Three strands of high tensile wire charged with a good energizer seem to work well to keep the hogs and sheep from creep grazing into the next paddock. My preference is at least a twelve joules energizer. I have only grazed finishing hogs in this mix, so I'm not sure how well this will work with weaners under 150 lbs. Also, hogs require a ring in the router to prevent damage to the grass. It is something a fellow who grew up in Scandinavia taught me, and it is very simple and effective.

Your water trough should be low and narrow, with a step if possible. This will allow all classes of animals to easily water, with little risk of getting trapped in the trough. Hogs don't seem to make a mess because there are animals coming to drink continually throughout the day, so they can't wallow beside the trough. Often there will be two or three species drinking side by side.

Moving the MOB

In a perfect world, mature animals should be trained to act as a MOB prior to them giving birth. Earlier I discussed how to do this. If the MOB isn't trained prior to birthing, you still use the same technique, just tone it down a bit. Be a baby wolf instead of a full-grown wolf!

If you are consistent, each individual learns its safe place is within the group. Once they are trained, a simple verbal scolding is all it takes to have a wandering individual turn and head back to the MOB. I know it sounds like a tall tale, but a loud "Get back to the MOB!" is all that is required. These techniques have worked with all the species I have handled.

Grass Management

Like I stated earlier, I initially put different species together to save labour. It is much less work to move one group than to move two or three. Further, you save on maintaining infrastructure for each group. The real benefit of a multi-species MOB, though, is what it does for the grass and soil.

Each animal brings something beneficial to the soil. Each species "massages" the soil in a unique fashion and has a preferred type of forage. Horses are a great benefit because of their large hoof. They do a lot of trampling that lays down a nice mat of ground cover.

I discovered this when we grazed around sixty bucking horses on a monoculture of alfalfa. After only two rotations, the horses had laid down a beautiful mat on the bare ground between the plants. They did this because of their large plate-like hooves and the fact that they don't have to ruminate, so they are constantly moving.

Sheep and cattle work well together because their palates do not completely overlap. Sheep will seek out more weedy forage as well as any saplings growing in the sward. I have noticed, though, that throughout the summer, calves tend to eat the same things the sheep like because they tend to hang out with the sheep a lot.

No matter if your MOB has one species or more than one species, you must follow a couple of grazing principles. First, your rotation must be in tune with how fast the grass is growing. I made this point already, but in my opinion, a grazing plan is critical! I wrote my first grazing plan in 2000. After more than twenty years, I still rely on a grazing plan. With a plan, you can see how often you need to be moving before you go out to check the grass. It has been my experience that the "grazier's eye" cannot always be trusted.

To illustrate a time when my "grazier's eye" was off, I will relate an amazing event. At one ranch I worked, there was a paddock that had been under irrigation for about fifteen years and only grazed with sheep. The stand was mainly short fescue from over grazing. When we came around to graze with the MOB, I could see that the grass was not ready.

Consulting my grazing plan, I discovered we were already at 38 days of recovery. At that many days of recovery, the grass was not going to grow any more. It was just going to get old! So with some trepidation, I moved the MOB (sheep, cattle, horses, and hogs) into the paddock for half a day. Eight days later, from a distance, I could see the paddock was lush and green. Upon closer inspection, that same short fescue paddock was now covered with a variety of grasses about a foot high! I couldn't believe my eyes, yet there it was. Once again, I had confirmed the importance of the grazing plan and the benefit of a multi-species MOB.

The second grazing concept is the grazing period. When grass is growing fast, in three days at certain latitudes there is enough regrowth that the plant can be bitten again. Further, the new growth will be sought out because it is like candy to an herbivore. By leaving your animals in a paddock longer than three days at this time, the plants are being overgrazed and therefore weakened. To me, this is as serious as injuring the animals. The only difference is that we can easily see an injured animal. We can't readily see an injured plant.

Often, I will see paddocks with a huge amount of grass that has not been grazed or trampled. Early in my career I would leave the animals longer than planned, but the rotation became too long, and the paddocks ahead became mature and poor quality. You cannot catch up.

Several years ago, I decided to just keep following the plan and not worry about the abundance of grass left undisturbed. I discovered that the animals topped most of the plants, which kept them in a vegetative state. Certainly there are some plants that do go to seed. However, overall, the stand remains high quality, which is very beneficial when stockpiling grass for winter.

I know I have made these last points already, and I will keep making them because they are very important!

I mentioned earlier the idea of "massaging" the soil. When you have your animals acting as one unit, they tend to graze as a group; not as tight as strip grazing or ultra-high-density grazing, but close enough to have an impact on the soil. I was taught by our Holistic Management instructors that forty animals per acre are enough to elicit some herd effect. My observations would support that rule.

Strip grazing and high-density grazing are tools for brushing or for significant soil disturbance. These are tools and not to be used as a style of grazing. Again, in my experience, when a group of animals is trained to behave as a single entity, they have enough impact to improve both the grass and the soil.

The real benefits of multi-species grazing come from the animals being grazed together. The same benefits are not realized when they are grazed separately on the same piece of land. Certainly there are logistics to overcome. Until hogs are around 150 lbs, I do not add them to the mix because they require grain up to that point. I have not been able to figure out how to keep the sheep from getting into the self-feeder and eating all the grain.

Horses tend to cause casualties to newborns because of their cavorting nature. Hence, they are not in the MOB until after the first cycle of birthing. To finish lambs in four and a half months and hogs in seven months, the stand must be high in legumes. I prefer red and alsike clover, as well as cicer milkvetch over alfalfa. There is much less chance of bloat and the other legumes are much more palatable than alfalfa, so can be grazed just like grass.

I have not been able to run multiple species every year because of marketing issues, and at times I worked for someone who chose not to. However, the benefits keep me trying to get as close as I can to my dream MOB of cattle, sheep, hogs, and horses. Sometimes all it takes is some imagination and some courage to try.

CHAPTER 11

Economics of Skilled Animal Handling

For a couple years I worked on a ranch in chinook country. I love chinooks because the grass stockpiled during the summer and fall becomes bare and we can go back to grazing. The problem with chinooks, however, is that there is a lot of water lying around that turns to ice when the temperature drops again. Travel becomes treacherous for human and beast.

One day, I wanted to move a herd of 353 dry cows back to the bale grazing paddock because a significant blizzard was in the forecast. No big deal, except for the ice! The herd had to travel 1.5 miles across a minefield of ice and they were not too keen to leave the safe paddock they were in. That meant they would not follow me past the gate when I blew the whistle to which they were well trained. No problem, I would just drive them to the bale grazing.

Initiating movement for the drive was easy. Unfortunately, they started to circle past the gate. I carefully ran to the other side of the herd and initiated movement in the opposite direction. The idea was to get opposing momentum that would meet at the gate. The meeting of the two forces would then push some cows through the gate because there was nowhere else to go. On the second attempt, the herd popped through the gate and down the temporary alley.

The next hurdle was getting them over a large patch of ice and into the next paddock. The herd stalled at the ice as anticipated, so I had to create some energy without scattering the cows. To create energy, I started to holler, "Hup! Hup!" which was my signal to start moving—the herd had been trained to that sound in the summer. I then ran into the back of the herd. As soon as I had some movement, I backed off. I continued this action again and again. After five tries, the first cows tentatively walked across the patch of ice. There was only room to go single file, so I just stood back and waited for the entire herd to cross.

Once the herd was into the next paddock, I patted myself on the back because the rest of the move was going to be a cakewalk. I went back to the front of the herd and blew my whistle. Like the well trained animals they were, the herd immediately started bawling and following me. I was once again leading the parade!

Funny how things don't always go as planned, though! Just before leaving the paddock to cross the county road, there was another patch of ice. I wasn't even concerned about it because it was tiny compared to the big patch the herd had already crossed. However, to the cows, it was pretty big!

The herd stalled, then turned around and headed back from whence they came. With cat-like reflexes and cheetah speed, I ran around the herd and stopped their retreat. Once I had them stopped, I began a zigzag pattern and got them moving toward the gate again.

When the lead cows stalled at the ice, I sped up the zigzag and put the rear cows into a trot. The momentum of their trot pushed the lead cows over the ice, across the road, and into the next paddock. From there, the cows knew where the bale grazing paddock was, and walked the last half mile without any more assistance.

The reason I am sharing this episode is not to point out specific handling techniques I used. I share it because there was a real economic consequence if I was not able to get the herd back to the planned bale grazing paddock.

At that ranch, we did not put up any feed. We stockpiled grass and bought hay. The winter hay supply was delivered in September. We had planned out where to bale graze, how many bales would probably be used, how many bales to put in each row, and then the bales had been set out right off the truck. This way, the yardage cost was only $0.10 per head per day. This is consistent with analysis done by the Saskatchewan Ministry of Agriculture.

When using hay from the hay yard, we had to feed the herd every day. We rolled out bales with an older model 95-horsepower tractor, and the yardage cost was $0.45 per head per day. If I was not able to get the herd back to the bale grazing paddock, we would have been spending an extra $124 per day. Further, if we were unable to go back grazing for ten days, the ranch would incur $1,234 of needless expense.

I shared this story to make the point that when we are able to handle animals effectively, we have more options. In this case, the option was to save money on winter feed by having the ability to switch between bale grazing and stockpiled forage. When animal handling skills are lacking, then the only option is feeding every one or two days with a tractor. It is not uncommon to have your feeding costs jump by $0.90 or more /day when using equipment.

Earlier, I made the point that ranching should be simple, easy, and fun. Training animals to behave as a group and effective animal handling are part of the system to make this happen. I know our ego can get in the way of becoming better at something, particularly when it comes to animal handling. Most people think they are good at handling animals. I challenge you, though, to take a course even if you feel you are an expert. Normally a course does not cost much and who knows, you might learn a new trick.

CHAPTER 12

Guardian Dogs

Before I begin this chapter, let me make a disclaimer: I am not an expert on using guardian dogs. I have had experience with about 50 dogs and five different breeds. However, the dogs have taught me most of what I know and I have been blessed with four amazing guardians. This chapter is dedicated to Queen, Blizzard, Tug, and Ringo.

Like I said, the dogs I've known have taught me most of what I know about guardians. Our first dog, Queen, was a gift from a fellow who was retiring and felt guilty that he sold Queen's flock out from under her. That Pyrenees bitch set the standard for all other dogs that followed. We didn't have to do anything with Queen because she just went to work as soon as she came onto the place.

When we were planning on getting our first flock of sheep, we read a lot about guardians and visited numerous sheep farmers. A number of people were big fans of lamas and donkeys. When I asked more questions, they revealed they still had a few losses to predators. The people who had dogs reported no losses, which is why we settled on dogs.

One of the shepherds we interviewed told me to expect to go through a few dogs because the instincts had been bred out of a lot of dogs. He said that people had started keeping guardian dogs as pets, which watered down their ability to guard livestock. It was his belief that only 10% of dogs are any good as guardians. He also said not to train guardian dogs because that also detracts from their main function.

Like I said earlier, we were lucky that our first dog was a superstar and showed us what we should expect out of a good dog. Because of Queen, we learned that a good dog has several qualities.

Qualities of a good dog

The first quality of a good dog is the reason for having dogs in the first place; you should have no predation of livestock once your dogs are mature. A dog should reach that level of maturity once it is one and a half years old. Before that age, you may lose the odd animal because younger dogs can get faked out by predators and leave the herd or flock to chase an intruder while other predators come in. Once your dog gains experience, though, they learn that chasing a coyote or wolf across the pasture leaves the group they are supposed to be protecting exposed to attack.

A good dog doesn't get into trouble. They don't go to the neighbours to harass their dogs and they certainly don't chase your livestock. Good dogs don't fight with your other dogs, don't chew on livestock, and they don't intimidate people coming to visit. They know who is just there to visit and should be welcoming, which means they should be well socialized.

Being well socialized does not mean they hang out around the house. It means that they come when called and they welcome your attention when they are with the flock or herd. I know it is tempting to show new pups lots of love and I recommend doing that, just not at the house. Give them lots of love when they are out with your livestock. If you are consistent, it doesn't take long for a pup to learn that its happy place is with the animals you want guarded.

A good dog knows when to bark. It doesn't waste its energy barking just for the sake of barking. Every decent dog I have had has a deep bark to alert predators of its presence. That is the only time they bark. I call that its *working bark* and it gives me great comfort at night because I know they are being vigilant so I can sleep peacefully.

Another attribute I expect of a good dog is that it understands what I am doing. They don't get in the way when I'm moving stock from one paddock to another and are fine with me sorting off individual animals. I get a kick out of how dogs move along with the group even on longer moves. It's as if they are part of the MOB and they aren't running about playing with the other dogs.

Finally, a good dog knows that when I bring a new pup to the ranch, it is their job to train the pup. They will teach the new pup how to behave to my standards. They also show the pup love so that it can become a welcome member of our community.

Once again, I was fortunate that our first dog was so good because it helped me develop a *no excuses* attitude. It can be very easy to make excuses for a dog's poor behaviour, but if you do, your pocketbook will feel the pinch in the form of dead livestock. Remember, ranching should be simple, easy, and fun. Dealing with a dog's bad behaviour is not fun in my books!

When purchasing a new pup, I recommend buying your dog from a breeder who has the same *no excuses* attitude. Someone like that will have already weeded out poor genetics. They will also not need to give you a list of things to do to train your pup. I also recommend checking out the parents of your pup. If you get to the ranch or farm and the pups are being raised at the house, turn around and go home. The breeder should have to take you out to the pasture to see the pups. Those pups will already know that their place is with the livestock.

What are your needs?

Before you purchase a guardian dog, I recommend figuring out what you want from a dog. Ask your-self what predators are in your area? How large is the predator population? How many animals will you be running? How many acres will the dog have to guard?

The type of predators will dictate what breed you should get. Smaller breeds like Pyrenees and Maremmas are fine if coyotes are all you have to deal with. If you also deal with wolves, though, larger breeds like Kangal, Anatolian, and Akbash are more suitable.

I recommend having at least two dogs at all times. When you only have one dog, if something happens to that dog, you are left with no protection. I find two good dogs can handle up to 200 sheep. After that, you should have more dogs. I read that in years past when there were huge flocks in Eastern Europe (5-10 thousand) the herders used one dog for every 100 sheep, so that is the rule I abide by.

If your flock runs on large areas, training your sheep to *flock* will greatly enhance the effectiveness of your dogs. With sheep that don't *flock*, the dogs have a difficult time protecting numerous little groups. They can only cover so much ground at once. I have already explained how to train animals to behave as a group and it is very important in high predator areas. Take the time to train your animals.

Of course, you can also night pen your flock by bringing them in or setting up electric netting out on pasture, but that is a lot of labour.

You can find this and other videos at simplyranching.ca/video.

Another great way to make life easier for your dogs is to get your sheep bonded to your cattle. That way, all your livestock are one big happy family and the dogs can just patrol the perimeter of the MOB.

Types and breeds

In my limited experience, the behaviour a dog exhibits is an individual trait and less a breed trait. There are three behaviour styles guardian dogs can express. The first is what I call a *flocker*. Just like the title implies, it is where the dog is always with the flock. The second behaviour, I call a *patroller*. Those types of dogs have an area they patrol, marking their territory and barking to alert predators of their presence. The finale style of dog, is a hunter. Those dogs actively go out looking for predators. I have only seen Kangals, Anatolians, and one Akbash go hunting, but again my experience with dogs is limited.

I find what style a dog will become doesn't really show until the dog is about eleven months old. Many people want to see their dogs with the flock at all times and honestly, that is what I would like to see. However, I remind myself that the most important trait of a dog is that I don't have any dead livestock. If there is no predation, then things are working the way they should. It is the job of the dog to see that no predation occurs. How they do it is not my concern.

In the book *More Sheep, More Grass, More Cash* by Peter Schroedtler, the author relates his early experience with guardian dogs. Peter was not convinced his dogs were doing anything because it seemed as though the dogs were sleeping all the time; even though he hadn't lost a single sheep after his two dogs arrived.

To find out if the dogs were actually working, one evening he dressed up in camouflage and crawled out to the sheep pasture. Suddenly, he heard a deep 'woof' and looked up as a dog was flying through the air at him. He quickly hollered out to stop the attack and never again questioned the work ethic of his dogs! (As an aside, Peter's book is a great read about how he turned his money losing operation into a profitable, easy, and fun enterprise.)

If you have to contend with predators like wolves and bears, larger breeds may be required. The two breeds I have had experience with are Kangals and Anatolians which are usually hunters; as mentioned, I did have one Akbash that was a hunter, but I don't think that is the norm for the breed.

The issue with hunters is that they have a large territory. One rancher told me his Kangals had a territory of six miles. The ones I have had would patrol an area of 2-3 miles. I worked on only one ranch where we had to contend with coyotes, both types of bears, and occasionally wolves. At that ranch we ran Kangals, Anatolians, an Akbash, and three Merammas. I liked that mix because the hunters were our first line of defence on the borders of the ranch and the Merammas were the last line of defence because they stayed closer to the livestock.

Neighbours

Neighbours can be your biggest challenge when keeping guardian dogs. *Flocker* dogs are not an issue, but if you have a *patroller* or a *hunter*, neighbours can get pretty riled up when a strange dog crosses their property. Often it is best to visit your neighbours before you get any dogs. Often, all it takes is explaining what you are planning and how the dogs work. People that have livestock tend to be quite receptive to have guardian dogs around, particularly if they have predation on their livestock as well.

At the ranch where we had multiple species of predators, the neighbour who we were concerned might shoot our dogs, turned out to be a big fan. He calved a month and a half before we did and the dogs started guarding his cows as they were calving. He was puzzled at first because he would head out at daybreak to check his herd. Three of our dogs would be sitting on a hill watching his cows. He couldn't figure out why the dogs were watching his cows until he realized the ranch didn't lose a single calf to predation that spring. It was the first time in the ranch's history that they had no predation at calving time. Funny thing, though, he never offered to buy us any dog food!

In my experience, acreage owners and weekend ranchers don't have the same appreciation for guardian dogs as do full-time ranchers. I'm not sure why, but I have a feeling it is because their economic survival isn't dependent on what the ranch produces. A full explanation of the efficacy of guardian dogs doesn't always work and you may have to come up with a different strategy.

I know some people train their *patrollers* to stay on the property by chaining a plastic barrel or tire to a short chain attached to the dog's collar. Eventually the dog will get tired of dragging the barrel or tire around and decide to become a *flocker*. I know Greg Judy has used that method with good results. In my way of thinking, though, patrolling is in the dog's nature, so I prefer to just get another dog.

In my twenty-plus years dealing with guardian dogs and neighbours, there has only been one who shot our dogs. He admitted to shooting one dog, but I suspect he shot two others as well. All three were hunters, and they routinely passed through his property on their patrol. It was a significant blow to the protection of our flock because the ranch was in a high predator area.

Intelligence/Instinct

As I have mentioned, I have been dealing with guardian dogs for over twenty years. The more dogs I have seen, the more amazed I am by the intelligence/instincts of good dogs. I am in awe of how they can be incredibly loving and submissive one minute and the next minute they become killers.

I have never witnessed a dog kill a predator, but I have seen remnants of carcasses. On numerous occasions, though, I have seen a dog laying in a submissive pose while I'm giving them a good belly rub, then suddenly jump up and take off across the pasture after a coyote.

At one ranch, where I was production manager, I lived in town. When I arrived at the ranch each morning, I would see the dogs positioned across the pasture in different formations. Somehow they knew how to position them-selves in response to what I assume was a predator threat.

Once, there were two young grizzlies hanging around just to the north of the ranch house for several days. The owners had young children who had to get on the bus each morning. Three dogs positioned themselves on a hill between the bus and where the grizzlies were hanging out until the kids got on the bus. Once the bears moved on, the dogs no longer guarded that area.

On that same ranch, I introduced sheep to the operation. To build our flock of 250 ewes we bought sheep from several different places. I won't do that again because those sheep did not respect fences, and we only had one paddock that would contain them.

Although I was used to running sheep and cattle together, the ewes we bought were not bonded to cattle and would wander around the ranch willy-nilly. Since we only had the one paddock that could contain sheep, that is where we lambed. Unfortunately, the paddock was not near the cow herd.

The cows were calving at the same time as the ewes were lambing and the dogs stayed with the sheep, leaving the cows un-protected. One morning as I was checking the cow herd, I came upon a dead calf that looked like it was born perfectly healthy. After closer inspection, I noticed there was a large pool of blood under its head and the tongue and eyes were gone. The owner of the ranch was trained in predator verification and confirmed that the calf was killed by a bird (she is amazing at predator verification and could probably have a reality show called Predator CSI).

I hadn't seen any ravens in the area which have been known to kill calves and lambs. I did see four bald eagles though, but they only eat fish and carrion. I was flummoxed as to what was going on. Two days later, there was another dead calf killed by a bird. That afternoon, the mystery was solved when I saw a golden eagle flying around with four bald eagles following.

I had never dealt with birds of prey and started to panic because I didn't know how to deal with the golden eagle other than shooting it, which is very much illegal. After some thought, I realized I had to move the sheep to the cattle.

It was a slow process because the ewes were not used to moving yet and they also had new-born lambs at their side. Slowly though, over the course of two days, I was able to move the flock to the cow herd. In the meantime, the eagle killed one more calf and a heifer that had calving problems during the night.

Once the sheep, and hence the dogs, were near the cows, we didn't lose another animal. In fact, the next morning at daybreak I witness three dogs chasing the eagle across the pasture. The eagle must have realized the dogs were too much of a threat, because I never saw it again.

How did the dogs know to chase that bird away? I'm sure it was new to them, but somehow good dogs know what needs to be done and just do it. I did talk to one shepherd who lost a rash of lambs to ravens, and she excused her dogs because birds are not a normal predator. I can attest that good dogs can recognize predators no matter what form they come in. Again, no excuses for your dogs… period!

Side note: each predator leaves a signature. I was fortunate to have a boss who was a predator verification genius. The first time she demonstrated her prowess was when she examined a dead cow that was killed by a sow grizzly and her cub. There were no external signs on the cow except for two small scratches and only one small paw print in the mud.

When she skinned the hide back, the carcass was severely bruised along the back and sides. She also found several paw prints that weren't obvious to me. Those paw prints helped her surmise that it was a grizzly sow and cub. I was a bit sceptical, but the next morning her husband went out to see if the bears would return. He was able to get a picture of the two bears ambling across the pasture. Like I said, she could have her own reality show!

As for bird predators, their signature is blood running down the check of the animal and a big pool of blood under the head. Also, the eyes and tongue will be eaten. There is an artery near the eye that birds peck until it is punctured and the animal bleeds out. If you skin back the hide from the face, the muscles will be bruised from where the bird pecked, trying to puncture the artery.

My superstars

Before I share stories about a couple of my superstars, let me share the story of a friend at Athabasca and her superstar dogs. Actually, I don't know if her two dogs were superstars, but they sure improved her success running 300 ewes.

My friend swore by her lamas and donkeys. I shared our great success with our first dog, Queen, but she wasn't convinced. Then one spring she called me and asked if Queen was really as effective as I said. She thought she had been losing lambs each spring, but didn't really know because her ewes lambed out on pasture and she didn't see any carcasses of dead lambs. That spring, though, she couldn't deny it any longer because the losses were huge.

A week after our conversation, she found two eleven-month-old Merammas for sale and brought them home. As simple as that, her losses ended. Her weaning rate went from 0.7 to 1.8 the next spring. She was thrilled with her dogs and wondered why she hadn't gone to dogs sooner.

Queen: I have already mentioned that we lucked out getting Queen because she was a mature dog that was already a great guardian. When she was introduced to our small flock, she immediately jumped the fence in with the ewes. The ewes weren't too keen to see a dog, so Queen tilted her head and averted her eyes. It was just like I had read about in articles about guardian dogs. In short order, the ewes accepted her and she once again had a flock to watch over.

The first week we had Queen, my wife, Jan, was walking our son, Alex, to the bus for school. Queen was trotting ahead of them when suddenly she turned, gave her signature *woof,* and headed straight toward my wife and son. Jan thought they were going to be attacked by the dog, but Queen raced past them down into the field. Apparently, a coyote didn't get the memo that we had a guardian dog on the place.

The coyote started running away, then stopped to face Queen for a second. That was a fatal error because when it started to run again, Queen pounced, grabbed the coyote by the scruff and snapped its neck. Queen simply dropped the coyote and came trotting back to Jan doing her happy dance. She always did a happy dance when she thought she had done something good.

That was the only time we could confirm Queen had killed a coyote, but we never lost a single lamb or calf while she was with us.

When we started pasturing turkeys, we were concerned how Queen would respond to poultry. We brought her into the brooding house and let her watch the poults. She promptly jumped the partition into the group of baby turkeys. She then laid down and let the young birds peck her fur and even jump right on top of her. I think she may have regretted that move because when the turkeys were out on pasture, invariably they would find where she was sleeping and promptly jump on her as though she was a new found perch. Not once did we ever see Queen molest a turkey; the same couldn't be said for my border collies.

Once, when Jan and Alex were walking along a trail through the bush, Queen started whining and got very agitated. Jan didn't think much of it and they continued walking. Queen then knocked Alex down and continued whining. That's when Jan decided they should probably turn back. Just then a cow moose and her yearling calf walked onto the trail. From then on, we always paid attention to Queen when she began getting agitated. I know she should have been guarding the sheep, but sometimes people need protection too.

Blizzard: Blizzard was an eight-month-old Meramma bought from a shepherd with a *no excuses* attitude. That dog had nine lives. On the way to the ranch, where I was production manager, I also picked up twenty young ewes to add to our flock. About a half hour before getting to the ranch I had to slow down to make a sharp corner when Blizzard decided he had enough travelling and some-how jumped out through the narrow window of the trailer. Of course I didn't notice his escape until I opened the trailer door at the ranch. The back compartment was empty!

I was panicked because I was still new to the ranch and introducing sheep was my idea and now we had just spent $800 on a new pup and I didn't even make it back to the ranch with it. I didn't know where to begin looking for the dog, so as a Hail Mary I called the local radio station to report the dog missing. Would you believe that twenty minutes later I received a call from a woman in Dawson Creek, an hour and a half drive from the ranch? She had the dog.

The woman saw Blizzard sitting in the ditch right where I had slowed down for the curve and picked him up. She said she would drive half way and meet me. I was elated! When we met up, she led Blizzard out of her car and handed me the leash. Boy was I thankful. That gal must have been quite a dog lover because her back seat was filthy from hauling a grubby farm dog around and it was of no concern to her.

The next incident with Blizzard happened when he was eleven months old. One day Blizzard disappeared and was gone for a week. I assumed he had tangled with some wolves and met his demise. Then on the eighth day I saw a white dog come trotting across the pastures. It was Blizzard except he had a red collar on. I couldn't figure out how his collar went from brown to bright red until I examined him.

His neck was cut wide open in a perfect circle. Apparently the neighbour, the weekend rancher, had a bait pile just on the other side of our south fence. He had snares set for coyotes and Blizzard got caught in a snare. The rancher set the snares before he left for his week away at work and the snares didn't get checked for the week he was away. Luckily blizzard had the where with all to stop struggling once he realized there was no escape.

There was nothing the vet could do for the wound, but it healed up nicely within a month and within two months, you couldn't tell that anything had happened. Blizzard's next brush with death was from a bear or a cougar. He received a long gash on the inside of his hind leg which took multiple stitches. Luckily it was like someone had cut him with a scalpel, the wound was that perfect. Interestingly, that vet bill was more than the bill from semen testing 13 bulls. Go figure!

I have mentioned intelligence a few times, and Blizzard was incredibly intelligent. He trained new dogs with equal amounts of love and sternness. When one of the subordinate dogs, an Akbash, lost its collar, Blizzard stepped in to help. The Akbash was a big dog and when I was putting on a new collar he started dragging me across the yard. Blizzard turned on the *Killer Switch* and pinned the dog down. I thought he was going to tear that dog apart, but the Akbash just laid there in submission.

Once I had the collar in place, I attached a lead and told Blizzard, "That's good". Blizzard released the dog and it got up. I hadn't trained the dog to lead so I took the opportunity to do some training. After a couple steps the Akbash threw a fit. It was like I had a marlin on the end of the lead. Once again Blizzard pounced on the dog and pinned him down. After a few minutes I asked Blizzard to release him. This time we made it about twenty feet and the Akbash threw another fit. Blizzard pounced again, pinning the dog down. Once again I asked Blizzard to release him. Eureka! The dog was trained to lead and I walked him around the corral yard for the next twenty minutes, not once pulling on the lead.

The last dog we bought for that ranch was an eleven-month-old Meramma whose parents had been imported from Portugal. That dog would occasionally rough up sheep, but Blizzard sorted him out in short order. The last time he had to reprimand that dog was when we were loading out some ewes for sale. One ewe cut back and started running to the other end of the pen. The Meramma grabbed a hold and started shaking the ewe. I ran over and gave the dog a good scolding then returned to loading ewes.

As I walked away, Blizzard grabbed the Meramma by the scruff of the neck and gave him a good shaking. Right after Blizzard, the Akbash did the very same thing. The Meramma must have finally got the message because that was the last time it ever roughed up a sheep.

I stated at the start of this chapter I'm not an expert on guardian dogs; my dogs have taught me how to guard livestock, not the other way around. What I do know, though, is not to settle and not to make excuses for your dogs. Further, when you get a good dog, trust them to do their job, however they see fit. Guardian dogs have been bred for generations to guard livestock and know way more about that task than any human could hope to know.

Stockpiling for Winter

CHAPTER 13

Stockpiling with Less Labour

The largest expense for most livestock operations is winter feed. My goal is to plan a grazing rotation so there is high quality, high-volume forage available in the fall and winter. Therefore, if I am going to stockpile forage, I may as well stockpile good forage. To my mind, having lots of old moribund grass in November is a waste of time because livestock do not like eating it and will certainly not thrive on that type of feed.

Now that the goal has been established, how do I accomplish it with the least amount of labour and hassle possible? Or more simply put, how would a twelve-year-old rancher stockpile grass? Let me start at the beginning.

It is my experience that high-quality fall and winter grazing is created during the first rotation in the spring. Therefore, I like to *drift* into spring grazing. That means animals have access to feed and stockpiled forage at the same time. However, they keep moving from paddock to paddock similar to a summer rotation.

The animals let me know when they are finished eating hay or silage by leaving the feed untouched.

As the animals move through the paddocks, they "wake up" the grass in two ways. First, the hoof action draws out frost in the ground the same way as hoof action in the winter drives frost in. In my oilfield days, during spring break-up, we would keep all traffic off a particular trail out of the bush until the very end, because any traffic over that road would draw out the frost and it would become a muddy, near impassable bog. The second way animals "wake up" the grass is by friction when they tear it. I don't know exactly how it works, but that friction gets the grass going in the spring. (I have recently discovered that waking up the grass is actually a physiological process called compensatory photosynthesis). This means a person can get a jump on stockpiling.

At this latitude, where I have been grazing since 2000, a thirty-five-day rotation on the first graze is a pretty safe number to use. Of course, the conditions will dictate whether you need longer or shorter. As you go further south, your average may be closer to forty days, but if you go much longer than that, you end up with "dry cow feed" as your stockpiled forage. If your grass gets into Stage Three, you are screwed for the rest of the year! Not only is the quality terrible, but if your grass regrows after being grazed in Stage Three, it will only produce around 50 percent of what would have been possible if grazed in late Stage Two.

At this time of year, *the grass need only be clipped*! Now is not the time to MOB graze and put down a heavy mat of vegetation on the soil. Back when I was concentrating on the grass, I ended up staying in paddocks until there was a nice even graze. This created a recovery period that was much too long and gave me Stage Three forage for the rest of the year, including Fall/Winter.

By about 2010, I stopped focusing on the grass during the first rotation and focused on my grazing chart instead.

I attended a MOB Grazing course at Greg Judy's place, co-presented with Ian Mitchell-Innes from South Africa. Ian explained that paddocks should be long and narrow because animals tend to walk to the end of the paddock before they start grazing. As they walk, they do a lot of trampling.

I have noticed that animals do walk to the end of the paddock when the paddocks are long and narrow. They graze as they walk, clipping the grass and trampling as they go. I have also noticed that when animals start behaving like a cohesive group, they create a MOB without the need for multiple daily moves. When that happens, you get trampling without the labour. This also creates competition, which alleviates a lot of selective grazing. (I explained- how to create a MOB in the section on animal behaviour.)

Now days I give animals the whole paddock. They will walk the entire area nipping only the very tips of plants. Many times I have been tempted to leave them one more day because they barely touched the sward. However, when I do a good walk through, I can see little bites on almost all the plants. When grazing sheep, you might have to look a bit harder!

If you are unable to drift into grazing, you still need to get out early. As I mentioned earlier, Jim Gerrish explains very well how to do that. The important thing to remember is that getting out early is critical for grazing in the sweet spot.

Since grass growth in spring is so rapid, often we are faced with too much grass. This is a great situation to have, but the grass still needs to be clipped. There is a debate that during this time making hay or mowing is a great way of controlling the spring flush. Both of these methods require the use of equipment, which means more expense. To handle the spring flush, I like to use animals instead of equipment.

Two ways to handle this situation are to break up your animals into two groups, or have a second group of animals that will leave the property the first week of August. Both methods create more labour because the second group must be moved separately, have a separate water system, and use another mineral source. This goes against the idea of decreasing labour. However, I have not been able to figure out another way of getting all the grass clipped without the use of equipment.

As I mentioned, there are two ways to have two groups. One method is to have a group that will leave the property the first week of August. This could be a group of yearlings that will be sold or sent to a feedlot at that time. The problem with this plan is the price may not be optimal for selling or it may not fit in with the management plans of the feedlot. Therefore, I prefer a method where you have more control.

If you are only grazing yearlings it can work well because you just have to split the group into two, then combine them the first week of August. If you are only grazing pairs (ewes or cows), then the same technique can be used. It gets tricky if you are grazing both classes of animals, though. I don't have a foolproof way of dealing with that situation.

So that is how I handle the first rotation. If the growing conditions are good, then the second rotation is the same as the first! Clip and go! Clip and go!… Until August 7.

After August 7, at this latitude, there are not enough daylight hours to produce abundant forage before a killing frost, so I slow down the rotation drastically. By this date, stockpiling paddocks have been determined for both Fall/Winter grazing and for calving. These paddocks are set aside and growth is allowed to accumulate.

Please note that August 7 is the date for Athabasca, Alberta. It will be a different date as you go further south because of different daylight hours and different killing frost dates.

As I have stated, I like to combine herds at this time of year. When herds are combined, the number of paddocks being grazed at any one time is cut in half. It also means there are more paddocks being rested. Now is also the time to strip graze, graze road allowances, graze riparian areas, or graze any other grass you can find. Each year, the more closely I have followed this method, the better the stockpile for winter grazing. Further, the more confident I become with the method, the better the results.

Here is my economic thinking on stockpiled grass, which to me is the whole reason to stockpile forage!

One year, our cost to feed a dry cow was $2.10 per day because of high hay prices. We kept 800 dry pregnant cows and 470 bred heifers on the ranch. This means that, for every day we used stockpiled grass, the ranch saved $1,680 in bought feed just for the cow herd. That did not include the days we grazed the 470 replacement heifers. Since we grazed sixty extra days, the ranch saved $100,800 on bought feed for the cow herd. Don't you agree it pays to stockpile high quality, high volume forage?

To verify my results, I get a feed analysis on stockpiled grass. It is like my report card. If the TDN (Total Digestible Nutrients) is greater than 60 percent, protein is 12 percent or higher, and volume is 150 stock days per acre or more, I get a gold star!

A final thought on stockpiling forage: I should mention that I only strip graze during the dormant season. In my experience, the growing season is mainly for clipping grass. Yes, there are times when MOB Grazing the animals is required. However, if a person is serious about stockpiling forage, MOB Grazing is the exception, not the rule. And if a twelve-year-old were ranching, they would rather be at the lake than moving cattle five times a day!

Here is how I stockpile in point form:

- Get out early. Drift into grazing if possible.
- Clip and go. Use your grazing plan as a guide, not the grass.
- If conditions remain good, Clip & Go.
- Slow down your rotation and combine herds by August 7. These dates will change depending on your latitude.
- Strip graze when there is snow on the ground to get better utilization.
- Test your grass for feed quality. Successful stockpile has a TDN greater than 60 percent and protein better than 12 percent.
- Don't bother stockpiling paddocks with a volume less than 50 SDA.

Stockpiling Grass Gives You Flexibility

From 2015 to 2020, I was Production Manager & Ranch Manager for a couple of ranches. My employers' priorities and mine didn't always mesh. I was focused on grass production, whereas they were focused on mundane things like cash flow, profitability, and keeping the ranch economically viable. Go figure!

However, since I had high-quality, high-volume stockpiled grass, I was able to be flexible when market conditions changed.

A stockpiled bush paddock.

Stockpiled grass after a snow then melt, in an open paddock.

Let me share a few examples of how this played out and could only play out successfully because of methodically stockpiling good grass.

One fall, we had a small group of bottom-end yearlings that were supposed to be marketed during the third week of September. However, the market was weak at that time. Since we had ample, good quality forage, we could keep the yearlings for another six weeks. The price rebounded, and the animals gained two pounds per day (after shrink) in that period. The owner was pretty happy.

The next fall, after I moved to another ranch, a similar situation arose on a larger scale. At that ranch, the owners fed the yearlings into a feedlot they owned. The initial plan was to ship the herd of 500 feeder heifers to the feedlot in the last week of September. The herd of 600 replacement heifers were scheduled to go at the end of October. Another 350 dry cows were to arrive at the ranch in mid-October. Finally, stockpiled grass was required for 800 cows to calve on the next May. "No problem!" I said, because we had lots of grass. Then plans changed. It's a good thing I always write my plans in pencil!

Apparently, the price for fat cattle was increasing every week. The owner was holding on to the fats as long as possible to capture the price rise. That meant there was no room in the feedlot until the fats were sold. Suddenly I was charged with keeping 950 dry cows and 1100 yearlings out on grass as long as possible! Compound that with the price of hay (the cost to bale graze a dry cow was over $2 per day) and things could have been pretty scary. However, since we had good quality, high-volume stockpiled forage, all I had to do was erase the old plan and pencil in a new one. If we had had poor-quality forage, grazing 1100 yearlings successfully on top of the dry cows would have been quite a challenge.

Incredibly Awesome Anecdote

On a tour of a feedlot, I noticed the newly weaned calves. They were out on beautiful, lush, green forage. Those calves looked happy and content.

The owners had planted a polyculture for silage and kept the regrowth for the calves. The calves were eating a mix of annual and perennial ryegrass, turnips, and a variety of other plants. They were doing something they knew how to do: graze.

After a few weeks on the stockpiled forage, the calves were introduced to silage and grain while still out on pasture. The owners had a great deal of success in terms of increased health and low death loss with their protocol. So much so that they trenched water lines out to the paddocks surrounding the feedlot so the animals could stay out grazing when the weather turned cold.

The take away here is that when there is good quality forage, your ways of feeding livestock can be highly flexible.

SWATH GRAZING

Swath grazing may not be something a 12/80-year-old can do every winter. There have been winters when the snow has been up to my thighs and it is very difficult putting up electric fence when you are tromping through deep snow in snow shoes! Of course, I have also used a snowmobile and that hasn't been too bad. So maybe a 12/80-year-old can still swath graze in deep snow.

I include this chapter because I have done a lot of swath grazing, and if done well, it can be far less expensive than other feeding methods. I have swathed grazed dry cows for as little as $0.56 per day, including yardage. So, as you can see, the upside is huge. However, there are some challenges.

The number one hazard people bring up is too much snow. I'm not sure where the limit is for what cows can dig through, but it is certainly more than we think! The winter of 2006 was a big snow year for our area. There were dry cows digging through three feet of snow to get at the oat swaths. We did not take any pictures, unfortunately, because I know it is hard to believe unless you see it yourself.

It has been my experience that as long as there is plenty of good quality feed on the ground, the animals will dig down to find it. This has only ever been a problem when the swaths got iced over. However, even if that happens, there is a solution.

I will break this chapter down into three sections: seeding and swathing, fencing, and operations.

Seeding and Swathing

Research by Baron et al. at the Lacombe Research Station found that if oats and barley are seeded later than June 7, production is 30 percent less than if seeded prior. Elgar Grinde, a rancher near Vegreville, Alberta with over thirty years of swath grazing experience, suggests that if those grains are seeded later than June 15, production actually drops to 50 percent less than if seeded in May.

The Lacombe research clearly demonstrated that, "for maximum whole-plant yield barley should be planted as early as possible, while triticale and oat should be planted approximately

31 May. However, yield maxima may not coincide with the target harvest date for swath grazing." Further, "low yield and beef carrying capacity [is] associated with [sic] extremely late planting date."

The target harvest date chosen in that trial was sometime in September. The researchers picked that date because they were worried about mould and other forage deterioration due to significant moisture in the fall. This is the piece of the research people focus on, and it is why most crops put in for swath grazing are seeded late June or early July.

Elgar is my go-to guy when it comes to swath grazing because he has so much experience. He challenged the authors of that research because it has not been his experience that mould and nutritional deterioration are an issue in a wet fall. One of the authors confirmed they did not have data to back up the claim, they just assumed it would be a problem.

I went to visit Elgar once in late February. We toured his oat and barley swaths. When I turned the swaths, they were still nice and green with very little mould. In fact, there was more mould in most cow hay than in those swaths. This was after being swathed the end of July and receiving nine inches of rain before freeze up.

There is some research that shows the deterioration of nutritional quality when small grains in swaths receive significant moisture (Arvid, 2004). That may be the case when the grain is swathed in a later stage of maturity. However, my experience with oats has been different when they are swathed in the milk stage.

In 2005 and 2006, we were able to purchase a salvage crop of oats each year. Both crops had been put in late with the purpose of making green feed. Both were swathed in the milk stage, but because of early snow both years, the farmers could not bale the feed. The snow melted, then we received more snow and that melted, which is to say there was a fair bit of moisture on the swaths.

The 2005 crop was tested in February. That feed test came back with a Crude Protein of 14.7 percent and a TDN of 72.98 percent. The 2006 crop was tested in January with a CP of 14.6 percent and a TDN of 71.09 percent.

Swathing early is vital to having winter forage that is good for all classes of livestock. Elgar starts swathing as soon as there is any water in the heads, because he has old equipment. He knows there will be breakdowns and by the time he finishes swathing, the grain will be in the milk stage. Research done at the University of Michigan on oat and peas as forage also used the milk stage of oats as the signal to swath.

After the crop is swathed, curing continues in the swath. If not swathed early enough, the plant continues to cure and you end up with grain and straw. In this scenario, when the animals are put on the swaths in the winter, they go through, pick out the grain, and leave most of the straw. It makes for a mess in the spring and you get lower utilization because the straw is wasted.

Marty Lawrence, a ranch manager near Stettler, Alberta, started planting a polyculture for his swath grazing. His mix contained oats, barley, wheat, peas, and Italian ryegrass. His results have been amazing! Not only does he get plenty of regrowth after he swaths, the utilization is exceptional. The winter of 2020–21, he grazed 450 dry cows on his polyculture. It produced 394SDA and his cost was $0.67 per head per day. Now that's going to another level in my books!

When you do start swathing, hire a swather with a twenty-five to thirty-foot header. The large swath makes it easier for the animals to find the rows and there are fewer problems with snow and ice. If the feed is of good quality, both stems and heads, there will be little litter left over. However, if the feed is just grain and straw, you will have issues during spring seeding.

Incredibly Awesome Anecdote

In the spring of 2006, we received about ninety bred cows. They were thin, about to calve, and not the kind of animals a custom grazier wants to be getting in mid-March, but they were cheap. We unloaded them onto the oat crop I mentioned earlier, and they started calving after three days. When we moved the group to stockpiled grass at the end of April, they were a different-looking bunch! I bet they had gained one hundred pounds each without any supplement other than a mineral pack. Let me stress once again, swathing early gives a person a lot of flexibility.

Fencing

It is best to plan out your swath grazing before you begin swathing. I like to break the field into strips no wider than two hundred metres (six hundred feet). That is about ten posts spaced at twenty paces. Once you have a plan, swath in such a way that your temporary wires will run between the swaths and not over the swath. There is less chance of animals knocking over the posts that way.

After the crop is swathed, install semi-permanent fence to make your strips. For these, I like to use the same framework as I explained in my fencing template. Certainly, you can get away with a much simpler design. However, when it is -30 degrees Celsius and you are up to your behind in snow, it's not much fun trying to shore up a fence post because a moose or elk went through and broke it off. It's also not much fun trying to gather three or four cows from the next strip because they figured out how to sneak under or jump over the wire—just my opinion.

Operations

Swath grazing fits well in a winter feeding system when you are flexible. There are times when it would be better to have your animals on stockpiled grass or bale grazing rather than on swaths. When the ground is not yet frozen, your animals would be better off grazing stockpiled grass. However, once the ground is frozen, there are a few tricks to the trade to dole out swaths effectively.

First off, you must have a well-grounded, high-joule energizer. In high snow areas, not less than an eight-joule energizer is recommended. To control consumption, I like using a geared reel with orange turbo wire or equivalent. The orange wire shows up much better for wildlife.

For posts, I prefer ones with a single spike that have some flexibility when cold and have wire holders that are easy to access at various heights. To put the posts in, my preferred method is with a cordless drill and a twelve-inch-long masonry bit.

The long bit allows you to drill a hole without gumming up the drill chuck, and you don't have to bend down as far. I like the masonry bit because it lasts longer than regular bits when drilling through frozen ground. I have not used free-standing posts, but my friend Marty has a design he says is going to revolutionize the industry!

Your first two to three moves will be an experiment to figure out how many swaths to give. As a friend, let me recommend putting up the wire for the next move ahead of time. If the herd goes through the first wire, there is a good chance the second wire will hold them from walking the whole length of the field.

When there is plenty of snow, I like to budget three days' worth of feed. If there is very little snow or the swaths are mature, you should probably go to one to two days. In the case of mature swaths, the animals will eat most of the grain the first day and be left with only straw for the second day. They will do fine with one day of straw, but they will lose condition if they are getting two days of straw.

In the case of little to no snow, or if the ground is not yet frozen, I move every day. If I don't, there tends to be a lot more wastage because the feed gets fouled by trampling and manure. You can also set up the temporary wire so the cows have to graze under the wire. It is a great technique I use when feeding pellets to a group of bulls or yearlings as well.

When I have grazed pairs or sheep and cows, it has not concerned me if the calves and sheep go ahead. This drives some people crazy, so they use two wires to keep the calves with the cows. Running two wires has not worked well for me, so I just say, "Screw it!" Calves should be gaining as much as they can, so they should be getting the best feed. Sheep don't seem to make much difference to total utilization, so once again I am fine with them going ahead.

If you encounter a big snow year, and I'm talking three to four feet, run some horses with your cows. I have read that running 10 percent horses with your cow herd is all that is required to open up stockpiled grass. Let me assure you, the same works for opening up swaths.

If you or your neighbours do not have horses, it might be time to go to bales. However, before you do that, let the cows decide.

If your swaths are of high quality, the cows will go through a lot of snow to get to them. They will let you know by not going out grazing for over a day.

When you encounter thaw-freeze conditions for a few days, the tops of the swaths will become frozen. This is a serious problem if you have small swaths. To rectify the situation, once again add horses to the mix. Even if you have to graze them for free, you will save money over going to another feed source. Freezing down has not been a problem with large swaths. That is why I only hire a swather with a twenty-five- to thirty-foot header.

One final thought: plan for a direct route to water. Cows are not rats and are not that great at solving mazes. If there is a chance they will make a wrong turn and get stranded on the wrong side of a fence, you might have to make a trail for them to follow. You could also lead them to water the first couple of times, so they make their own path.

When I first started swath grazing, I used snow as the only water source. There is research from the 1980s that found dry cows did fine getting water from snow. It has been my experience, though, that about 10 to 15 percent of the herd does not do well on snow alone. This is exacerbated when the feed is less than 58 percent TDN.

If you are relying on snow, fresh snow is the best. However, in our area, there tends to be a period in late January when we don't get much snow. I have been caught more than once when I encountered these conditions and didn't have any water backup. Now, I just plan for water right from the start. It's not fun scrambling to set up water when the snow has disappeared or becomes inaccessible.

My second final thought: swath grazing should be simple, easy, and fun. For me, that is going out on my horse, pulling a toboggan with temporary posts, a cordless drill, an extra battery, and a spooler. I enjoy getting a good sweat going every few days because I'm not that active in the winter. For you, this may sound like torture. Remember, Ranching Like a 12-Year-Old is not a specific way to ranch. It is a state of mind. A way of thinking so that *your* ranch becomes simple, easy, and fun!

Incredibly Awesome Anecdote

If you live in an area with lots of wildlife, they may become a problem when you have high-quality feed sitting out in a field. A 3D fence may be a viable option to keep your swaths safe from herds of elk and deer.

When I was production manager of a ranch in high elk country, we had two hay yards. One had traditional elk fence and the other had 3D fencing. The 3D fence was just a single strand of high tensile wire installed four feet outside a four-strand barbed wire fence. In the eighteen years they had the 3D fence, only one deer ever figured out how to get into the hay yard.

This solution was much cheaper than traditional elk fence. That being said, I have not yet met anyone who has used 3D fence for protecting swath grazing.

Business & Economics

CHAPTER 14

Is Your Grass Making Money?

Once you get your grazing organized, you will see a lot of benefits. You will have more grass and more water, and then birds, bugs, and wildlife will return. These are pretty cool benefits! Many would call this sustainable agriculture. There is another aspect of sustainability, though, that doesn't get much attention and that is economic sustainability.

Ranching for Profit has a program after the initial course called Executive Link. It is for people who really want to make improvements to their operation and be held accountable. At most meetings, an industry leader is brought in to share some of what they are doing on their operation. At one meeting, Kit Pharo from the Pharo Cattle Company spoke about his grass-based cattle genetics and his ideas on how to raise good seed stock.

What he had to say was quite interesting. However, the most interesting and profound thing he said was that for agriculture to be sustainable, it must first be economically sustainable. For a ranch to continue on, young people must want to take the operation into the future. Young people do not want to take over an operation that isn't profitable, because it is not fun to live in poverty.

Kit's revelation got me to thinking about the economics of our grass. I wanted to know empirically how much each paddock was producing from an economic point of view. Over the course of several years, I developed a spreadsheet to give me that information, and I even added colour! Pretty cool!

A lot of agriculture production is expressed as dollars per acre. When we were custom grazing, I liked to have our revenue and expenses explained in "per acre" terms because that is how our land rent was paid. When dealing with grass, as I have stated earlier, I use stock days per acre (SDA) because it is immediate and basic. You don't have to do any more calculations based on class of stock or length of time. If you recall, one stock day is the amount of forage a 1000-pound dry, pregnant cow will eat in one day, and that cow has a value of 1.0 standard animal unit (SAU).

As you change your class of animal, the SAU for that animal will change. If you have a 1300-pound dry, pregnant cow, her value is 1.23 SAU. A 1300-pound lactating cow with calf at foot has a value of 1.75 SAU (these figures are from Holistic Management's "Standard Animal Unit Tables").

We rarely graze only one animal at a time, so the SAU for the herd must also be calculated. If there are 80 cow/calf pairs in a group and the cows weigh around 1300 pounds each, the SAU for the herd is 140 SAU (80 pairs x 1.75 SAU = 140 SAU).

Once you have the herd's value, you then need to know the size of the paddock. If your paddock size is 20 acres and your herd grazes for one day, the herd of 80 pairs has taken 7 SDA (140 SAU / 20 acres = 7 SDA). If the herd grazed for three days in the same paddock, the SDA would be 21 SDA ((140 SAU x 3 days) / 20 acres = 21 SDA). The graze days and paddock size can be found on your grazing chart. The SAU of the herd has to be calculated separately as the herd size changes. Once I move the herd to another paddock, I write the graze days in pen so I can access the data quickly when I do my analysis at the end of the season.

Throughout the season, record the number of graze days and the herd size for each paddock. Normally, I get two rotations during the growing season & one graze during the dormant season at the latitude of Athabasca, Alberta. When it is time to do an economic analysis, I add up all the stock days taken from each paddock, then divide by the size of the paddock to get the SDAs taken that year. It is then time to turn that figure into dollars per acre.

When we custom-grazed, the rate in our area at that time was $1 per pair per day. Since I work in SDA, I had to convert that rate into dollars per stock day. Therefore, $1 per pair divided by 1.75 SAU per pair becomes $0.57 per stock day. I use custom grazing rates because custom grazing is a choice. You decide if you graze your own cattle or someone else's cattle. This means your grass is separate from your livestock, so even if you own your cattle, they still need to be charged for the grass they eat.

The following table is part of the actual grazing summary for one grazing cell. Let's look at Paddock 2. From my grazing chart, I calculated 3,582 stock days were taken from the paddock during the growing season and 4,800 stock days during the dormant season. Since the paddock is 40 acres, the total forage taken from the paddock in 2018 was 210 SDA ((3,582 SD + 4,800 SD) / 40 acres = 210 SD). In this example, 1 stock day was worth $0.58, so Paddock 2 made $122 per acre (210 SD x $0.58 = $122).

Now that I have this information, my analysis turns to which paddocks are contributing most to financial prosperity and which are not.

2018 Grazing Summary

Paddock	Acres	Growing		Dormant		Totals	
		Stock Days	SD per Acre	Stock Days	SD per Acre	SD per Acre	Dollars per Acre
1	25	3108	124	472	19	143	$83
2	40	3582	90	4800	120	210	$122
3	63	4000	63	2400	38	101	$59
4	24	2400	100	2400	100	200	$116

Comparing Stock Days per Acre

Paddock	2017	2018
1	43	143
2	81	210
3	51	102
4	68	200
Total	243	655

I like this type of analysis because it is unambiguous. I know Paddock 2 made $63 per acre more than Paddock 3. There's no speculating: it's in black and white and I can compare what happens from year to year. Further, I can try a pasture amendment in one paddock to see how much of a financial benefit it produces. The grazing summary lets me know if that amendment was worth the expense and if so, maybe it should be done on more acres. Conversely, maybe it wasn't worth the expense and we should try something else. I find that pretty powerful.

Note: Since I have been using SDAs since 2000 and all my data as well as my thinking is in SDAs, I have not converted the tables to cow days per acre (CDA). If you want to convert SDAs to CDAs, divide SDA by 1.23 if your cows average 1300 pounds. If your cows are heavier or lighter than 1300 pounds, then the 1.23 will have to be adjusted higher or lower. The table I use to find standard animal units is from Holistic Management.

CHAPTER 15

Pasture Rejuvenation

Improving your pastures begins with getting a handle on the grass. It is the first thing I discussed in the Grass Management section. Once you are using a grazing plan and you have your fencing and water installed, only then is it time to look at improving your pastures. I recommend doing things in this order because invariably it takes two to three years to get your grazing house in order. There is only so much time and usually only so much money to develop infrastructure. I have often thought that we were going to get all our fencing and water done in one year because there was lots of help and money. However, it just does not happen!

The season for fencing and water development is the same as the season for calving and grass management. To follow the grazing concepts, the animals have to keep moving. Usually this includes training the animals to move effectively. Also, when you are calving or lambing, that obviously takes precedence over infrastructure, so improving fencing and water gets done around your other duties. That being said, let's begin this section with the premise that you are into year four and have identified some pasture that requires more help than just grazing.

In my experience, forage begets forage. When you have plenty of forage in your pasture, by following a grazing plan that keeps the sward in the sweet spot, you grow more forage. It becomes a self-propagating system. I know there must be a limit to how much a pasture can produce. However, because my bar keeps getting raised the more I learn and see, I don't know what that limit is.

When a pasture gets beaten down, at a certain point there is not enough biological energy in the soil to get the system going. Somehow you have to get significant forage growing. I'm not talking about doubling forage production, because if you have bugger all and you double it, you still only have bugger all. You need to really pop forage production! When you have a pasture that is producing less than 40 stock days per acre (SDA) over the course of a season, the sward needs to increase by four to six times to really make a difference. I only know of two ways to get that kind of pop.

The first is with bale grazing. In the next couple chapters I will discuss how to bale graze, the economics of bale grazing, and why placing bales is more effective than rolling out bales for improving pasture.

The second method is with ammonium sulphate. I know this is a bit sacrilegious to regenerative agriculture. However, if your goal is to get your pastures cranking, then sometimes you have to do something significant.

I equate pasture rejuvenation to driving a car. You can have shiny mag wheels, polish and shine the exterior, know the names of all the engine parts, know the model and the year, know the history of the car, but if you don't have any gas you aren't going anywhere! Forage is the system's gas.

There is so much information we can learn about the land we manage. There are courses and agricultural extension agents to help us understand soil biology, our soil types, and the best plant varieties for our soil, and yet we end up with unproductive pastures.

I have talked to many people who believe they have great pastures. When I start asking questions, they can tell me so many interesting facts about their land and why good grass grows in some areas and not in others. Invariably they go on to tell me about how much better production they are getting than their neighbour. Their neighbour's pasture is not the metric they should be using. You don't want to be judging your performance against a poor frame of reference.

Your bar should be set at a point that is profitable and allows you to raise your family without off-farm support. As a bonus, once you hit that target, the system becomes self-propagating. Again, I don't know where the endpoint is because my bar keeps getting moved up the more I learn and see. I also do not know exactly where the starting point for a self-propagating system is either. However, it appears to be around 170 SDAs over the course of a grazing season.

To me, self-propagating is a system where you no longer require inputs for your pastures to improve; all that is required is your management. By management, I mean a well-executed grazing plan that respects the Recovery Period and the Graze Period.

Before we continue, let me address the idea held by many people that you only require grazing animals following a planned grazing system to rejuvenate pastures. Planned grazing is certainly required to heal a damaged system; however, it can only do so much. Certainly, if you go from nothing to something, that is a huge improvement. That does not make the land profitable. We can talk about warm and fuzzy topics like all the wild animals that have come back to the ranch and the diversity of plants in our pastures, but it doesn't make the grass profitable. We need our land growing lots of forage to be profitable.

Dr. Alan Williams has a great webcast called Grazing 101. I highly recommend watching his three-part series. In the webcast, he shows what can be done using only cattle to graze. In his example, he bought some land in Mississippi that was farmed out crop land. The land was over-run with weeds and in four years he turned it into an amazing pasture with no inputs. The pictures are amazing!

That drastic turnaround was achieved with only animals and no inputs, but that has not been my experience with worn-out hay and pasture land. I asked myself what I was missing: why was Alan Williams successful when I had not been? Then it struck me! He had forage in the form of weeds to work with that first year. Remember, forage is the gas for the grazing system. If we don't have any gas or very little gas, we can't go very far.

Thinking back, I realized I had a similar experience with a field of weeds in 2004. A neighbour I worked for had a ninety-acre field that had been under-seeded for pasture the year before. All that came up in 2004 was a bunch of weeds. There was lots of forage, the forage was just weeds. I suggested strip grazing rather than reseeding. It only took the cows about 30min to figure out what was going on and then they went to town grazing down the weeds. Two rotations were done that year and the next year a heavy, healthy stand of tame legumes and grasses came up.

We have to have something to work with and I believe we need to do whatever it takes to get some forage growing. Our first target should be 170SDA days per acre of forage over the grazing season. Once again, when a pasture gets to that level of forage production, it becomes a self-propagating system. Once that happens, who knows where it will end.

BALE GRAZING

Bale grazing is a cost-effective way to improve soil health. If you pair it with planned grazing, you may only have to do it once. Many people have told me that they don't bale graze because the next summer all they get are weeds and spots where the litter is too thick to let grass grow. When I do some investigation, they reveal they don't have a planned grazing system or they skip the bale grazing field completely with the livestock. These people then go to rolling out bales.

In my experience, the only real drawback of bale grazing is that you can only cover so much ground each season. I started bale grazing in 2004 and have used it to improve pasture and as an adjunct to grazing stockpiled grass and swath grazing ever since.

The key to making sure your grass stand is not choked out by litter is to put the bale grazing paddock into your rotation, starting on your first rotation. The other key is to do something so the animals are walking over the high litter areas. Usually all you require is to mob them up a little or put the mineral tubs in such a way that the animals will walk into those high litter areas.

*February 2020: These two pictures show before and
after in the same paddock in the same year.*

*July 2020: In this second picture, you can see there are still
a few bales left over from the bale grazing.*

Another complaint I hear is about all the waste with bale grazing. Every feeding system has waste. With bale grazing, there is about 16 percent waste when the bales are placed on end and, in a trial by the Lakeland Agricultural Research Association, about 9 percent when the bales are placed on the round. The following table shows hay wastage, comparing various feeding methods, from research done at the University of Alberta by Barry Yaremcio, research by Buskirk et al. 2003, and by Kallenbach 2000.

Feeding Method	Hay Wastage
Unrolled on snow	12%
Processed onto snow	19%
Bale grazed on side	9%
Bale grazed on end	16%
Ring feeders	6%

When calculating wastage, people tend to forget about the huge benefit you get from bale grazing as a pasture amendment. I will share some of that data a little later on. You do not get the same benefit from rolling out bales.

May 2017.

The same paddock in June 2018. There were no bales in the foreground, so the clover was not affected. There is a line of no clover where the bale grazing bales started. Legumes will disappear for two years after bale grazing. In this paddock, we also broadcasted ammonium sulphate (200 lbs per acre) on May 4, 2018 in a test strip on the left hand of the photo. I estimated a four to five-fold increase in volume of alsike clover. The clover on the right side is deceptive. From this angle, it appears to be similar in volume to the left side, when in reality it was thin, sparse, and about half the height. The leaves were about half the size of the leaves in the test strip.

BALE GRAZING VERSUS ROLLING OUT BALES

One day my buddy Steve Kenyon and I were discussing why concentrated bale grazing is so superior for pasture rejuvenation versus unrolling bales. Steve is the author of *Calendar of the Year-Round Grazier*, teaches a course on how to graze year-round, and has been bale grazing for more than twenty years. Steve is a pretty smart guy and I respect his opinion. I have been bale grazing for almost as long as Steve and we have come to the same conclusion; rolling out bales does not improve pasture the way concentrated bale grazing does. But why?

Steve believes it is the litter left on the soil after we bale graze that improves pasture. He feels that the litter is as important as the nutrients deposited. When we unroll bales, there are sufficient nutrients deposited, but not much residue is left to cover the soil. I can't say for sure that Steve is correct on this point, but I certainly do agree with him.

When I talk about bale grazing, I am always referring to concentrated bale grazing. Placing bales out every few days in a haphazard manner does not produce the same effect as setting up rows. It seems there must be a certain concentration before a pasture will cross a tipping point and really pop.

As I said earlier, "popping" to me means a four- to six-fold increase in grass production. It's kind of like if there was a very poor community of eight hundred people and you gave each person $10. With $10, each person could buy a goat or maybe a bicycle. However, if the individuals pooled their money, they could purchase equipment to pump water for irrigation. I see concentrated bale grazing the same way. By concentrating the nutrients and litter, we can do something much more significant than spots of high production throughout the pasture.

Concentrated bale grazing can completely change your pasture. As an example, in 2017, we had a fourteen-acre paddock that produced only 39SDA of grass. That winter, we set up bales in rows forty feet apart, with the bales twenty feet apart. During the grazing season of 2018, we harvested 196SDA from that paddock. That is a five-fold increase! When you put that kind of increase together with a carefully considered grazing plan, you are well on your way to creating an amazing grazing system.

Fly By Night Ranching								
2018 Grazing Summary $0.58/SD								
-	-	Growing		Dormant		Total	$	2017
Pad-dock	Acres	Stock Days	SD/A	Stock Days	SD/A	SD/A	$/ac	SD/A
12	14	592	42	2156	154	196	$113.85	39

Fly-by-Night Ranching: 2018 Grazing Summary

Stock Days per Acre in 2017:   39

Paddock: 12

Acres: 14

	Growing	Dormant	Total
Stock Days	592	2156	2748
Stock Days per Acre	42	154	196

This table demonstrates the significant financial benefit of bale grazing on worn-out pastures. The paddock was flash grazed only once during the growing season because we didn't have enough cattle for the grass, so I had to make some paddocks a priority over others. I ended up letting this grass mature, then grazed it in the dormant season.

Note: To calculate the financial benefit of bale grazing, convert the amount of forage harvested into dollars per acre based on the custom grazing rate for your area. For example, let's assume the custom grazing rate in your area is $1 per day per pair, where one pair has the equivalent standard animal unit (SAU) of 1.7SAU. You then divide $1 per day per pair by 1.7 SAU to get the value of one stock day of forage for your area.

After November 1, calculate what it costs to feed a cow for one day and use that value to figure out the cash value of one *stock day in the dormant season. I use November 1 as the cut-off date because I figure most people are usually feeding by that date. From the numbers in the table above, the final result is $113.85 per acre! In 2017, the gross revenue from that paddock was only $22.62/ac. It's hard to raise a family when you are grossing less than $23/ac.*

As I mentioned earlier, I started bale grazing in 2003. That year, when the swath grazing was done for the 180 cows we were custom grazing, I started rolling out bales with two of our saddle horses. My dad designed a contraption that looked like large ice tongs, so my input costs were pretty low. The yardage to feed that herd with horses was $0.25 per day per head. I don't think the horses were too impressed with me, but boy were they in good shape when spring arrived!

I know my yardage was $0.25 because a friend was bragging one day about how cheap he could feed with his bale truck. In fact, he figured he could feed cheaper than I could with my horses. He kind of got me angry, so I challenged him to a feeding duel!

We agreed on an hourly rate for the horses, his truck, and our labour. After the dust settled, his yardage was $0.35 per day, and mine with horses was $0.25 per day. Then I heard about this thing called bale grazing. When I calculated the yardage for bale grazing, it was only $0.10 per day. Boy, did my horses celebrate, because that was the last time they ever rolled out another bale!

Here is a table comparing the cost of rolling out bales versus bale grazing for a herd of one-hundred cows. I have valued the yardage using a tractor at $0.45 per day based on data from the Alberta Ministry of Agriculture a number of years ago.

Yardage Cost Comparison Feeding 100 Cows for 100 Days

Unrolling Bales	1 Day	100 Days
Tractor	$45	$4,500
Bale Truck	$35	$3,500
Horses	$25	$2,500
Bale Grazing	$10	$1,000

Note: for this example, I am only including the yardage differences between the methods. Each method incurs other costs associated with yardage calculation.

One final note: I budget 35 pounds per day for a 1350-pound cow, which is the same amount many people budget when they feed every day with a bale truck, tractor, or bale shredder.

WATCH:
Bale Grazing

You can find this and other videos at simplyranching.ca/video.

Incredibly Awesome Anecdote

In 2008, I bought hay for a herd of 550 dry cows. Since I was short of cash, I arranged for the farmer to place the bales on a pasture he was going to plow up. It saved me the cost of trucking because we just walked the herd to his place, and he received free pasture rejuvenation. To this day (2021), he has not plowed up that pasture.

AMMONIUM SULPHATE

The use of synthetic fertilizer goes against everything I was taught in Holistic Resource Management. In fact, I held true to that ideology even though I could see that my worn-out pastures had reached a plateau that was not profitable. Finally, I had to stop kidding myself and face the fact that what I was doing wasn't working. That's when I contacted an agronomist I knew from our local forage association.

The fellow was an independent agronomist who had quit working for the government, so he could make recommendations outside government-approved parameters. He had accomplished a lot of amazing results in the grain world, but he had never worked with pasture land. I quietly asked him to see what he could do on our land. I never even told my buddy Steve!

After the soil tests were analyzed, he had all kinds of rec-ommendations, because the soil was so depleted. Since cash was tight, I asked for the biggest bang for the buck. His answer was two hundred pounds of ammonium sulphate. He thought that we should help out the legumes (red and alsike clover) with the sulphur, and the clovers would help out the grass. The nitrogen was just a bonus.

Before spending a whole pile of money, we started on a small scale. The test plot was twenty acres. For the trial, we included ten pounds per acre of red clover, which wasn't required in hindsight because of the abundance of alsike clover. Costs included picking up the fertilizer from town (freight). Here is a breakdown of the total costs:

Seed and fertilizer	*$1960*
Freight (est.)	*$200 (4 hours x $50 per hour)*
Spreading	*$375 (5 hours x $75 per hour)*
Total	*$2,535, or $84.50 per acre*

I got the results I hoped for. As I mentioned earlier, pastures often require a four- to six-fold increase of forage to get going and become a self-propagating system, and that is the result we achieved. I am not happy with the cost per acre of the amendment, but it is a one-time expense and you can improve as many acres as you can cash flow. With ammonium sulfate you are not limited to how many acres you can improve, as you are with bale grazing.

Because of the price of ammonium sulphate, I keep looking for something less expensive. Every time I hear of an amendment with amazing results, I track down the end user or the actual trials. Each time I find the same thing; forage increased by 50 to 100 percent. That may sound like an amazing feat, but again, when you start with bugger all and you double it, you still have bugger-all! Just because something is better than it was doesn't mean it is good.

This is the same picture I showed earlier, demonstrating the effect of adding two hundred pounds per acre of ammonium sulphate. It was just a strip to see how the application would respond in a bale grazed paddock. The clover on the left side is healthy and robust compared to the right side. Again, it is a one-time application when used in conjunction with a planned grazing system. Ammonium sulphate should only be used when effective grazing does not produce desired results

A soil biologist once told me that "if you build a home they will come." He was talking about soil biology: soil biology requires a roof over its head (thatch or ground cover), food (root exudates), and air (good soil structure). If one of these is missing, soil biology cannot thrive. Growing abundant forage helps you satisfy the three pieces to the puzzle.

Significant plant biomass provides the cover required to protect the soil by covering up bare ground. When you have plenty of vegetation above ground, there is abundant material below ground in the form of roots. Abundant roots create spaces for air to permeate as well as build soil structure, which increases water holding capacity. The trick is to get abundant vegetation above ground.

As I have mentioned several times, our bar for a robust, productive pasture may be set too low. When we start out producing 39SDA, then add an amendment such as compost and the pasture produces 75SDA, we still do not have a self-propagating system. Again, it has been my experience that the tipping point for that type of system is around 170SDA. Just a note: once you get there, it doesn't take very long to get to 350SDA and beyond.

Before I leave ammonium sulphate, I want to mention that I will be trying Turf-Grade elemental sulphur to forgo the nitrogen portion of ammonium sulphate. Research has shown that nitrogen oxide (NO) is released from nitrogen fertilizer. Since we are in greenhouse gas emission storage times, the less we emit, the better. Further, research by S. S. Malhi, titled "Effectiveness of elemental S fertilizers on forage grasses," found that Turf-Grade elemental sulphur produced comparable results to sulphate forms of fertilizer. As of this writing, though, I have not tried it, so I am unable to relate my findings.

A final warning: do not start down the road of amendments until you learn how to use a grazing chart and plan out your grazing rotation. Until your grazing is under control, money spent on amendments is wasted!

CHAPTER 16

Animals as Landscapers

On every ranch, there are areas that require some form of modification. These areas may be weed infestations, capped soil, or brush encroachment. Using a tool we already have can be a cost effective way of correcting these issues. This tool is our livestock.

Kathy Voth, editor of the on-line magazine On Pasture, had another life before starting her magazine. Before On Pasture, Kathy developed a system to train cattle to eat a variety of weeds and brush. Through her method, she helped many ranchers train their livestock to eat plants that most people believed livestock wouldn't eat. She demonstrated that animals can do a much more effective job of controlling weeds than the various chemicals traditionally used in agriculture.

With Kathy Voth's program, you don't have to train all your animals. A small group of animals that get trained to eat targeted plants will train the rest of your animals once they are back with the larger group. I have personally seen the long-term results and can attest to the fact that her system works very well. After animals have been trained, those plants become part of the smorgasbord when livestock are out grazing. That means they will eat a little bit of a large variety of plants. What happens, though, when you are dealing with a concentrated infestation of weeds or brush? This chapter explains how I deal with those areas.

For areas that require a more focused approach, I rely mainly on animal impact. You may recall my example of the small herd of bison that occupied our worksite in the North West Territories. After they moved on, the entire site was covered with hoof prints.

To create animal impact, I try to create excitement and/or competition. This usually involves temporary electric fencing, but once animals are trained to act as a group, the necessity for electric fencing decreases. As I have mentioned, when animals behave as a group, they tend to stick together and move around the ranch as a MOB.

In this chapter, I will discuss how my thinking was developed and give several examples of how I have used livestock to alter different landscapes. Once you understand the principles of using livestock for this purpose, often it is just a matter of trying it out. You may not always achieve your desired outcome, but more often than not, the results will be significantly better than you could imagine. The benefit of not having to spend any money is certainly worth the effort of trying.

Weed/brush Managers

I have been greatly influenced by the research of Dr. Fred Provenza, who spent his academic career studying why animals eat what they eat. In his writings and lectures, I learned that what animals eat is not only influenced by taste but also by what they have been taught to eat. Young animals are taught by their mothers to eat certain foods and not eat others. They are also affected by what the mother eats while the baby is in utero.

This scenario works well and helps young animals grow and thrive. However, it also has a drawback in that a young animal's palate is only as large as the mother's palate. There are many nutritious foods that could be eaten, but are out of the mother's realm of experience, so offspring don't eat those other foods. Children experience that same situation as we learn to eat.

There are many foods that are seldom eaten in North America, yet are consumed regularly in other parts of the world. Chevron is a one example. Besides myself, I am the only person I know who was born in Canada and has eaten goat meat. In Nigeria, though, according to a Nigerian friend, goat is a delicacy. When celebrating special occasions, Chevron is the meat of choice. Even within North America, there are foods eaten by some people, yet others find repulsive. One such example is moose nose. A Cree friend raves about how his kokum prepares moose nose when he goes to visit. The first time he mentioned the dish, I instantly turned up my nose (no pun intended)!

Dr Provenza contends that the social aspect of palatability is just as strong as the actual flavour and nutritional value of food. This aspect of nutrition explains why some animals eat weeds and some won't. It comes down to what they have been exposed to.

I have already mentioned Kathy Voth and her training program. That training system came about after Kathy audited one of Dr Provenza's classes at Utah State University. After that class, Kathy questioned if what animals do not eat is a result of not being exposed to certain foods, rather than toxicity or flavour.

I can only imagine the scepticism she faced as she developed her Cows As Weed Managers protocol.

My first foray into weed management actually occurred before I became familiar with Dr Provenza's work. We had just bought land from my wife's parents. On the west end of the property there was a forty acre pasture that had been continuously grazed for many years. Within that pasture were several thistle patches; one was at least one acre in size. Putting my Holistic Management training to use, I placed the mineral tub in the middle of the thistle patch, expecting the herd to trample the thistles.

When we first started out we had no internal fences and were building two acre paddocks every two days for our forty cow/calf pairs. That meant the herd only had two days to affect the thistle patch. My expectation that the thistles would be trampled did not occur. The cattle did make trails to the mineral tub, just not much trampling. What did happen, though, was that as the cattle walked to the mineral tub they bit off the emerging flowers of the thistle plants.

I was pretty disappointed by the results. However, the next season there were less than 5% of the original thistle plants in that one acre paddock. It was like a Christmas miracle! That event really got my creative juices going, thinking about what else we could do with a mineral tub.

If you recall I took a Ranching for Profit course a few years after Holistic Management. The year I took the course was the same year the RFP conference was held in Calgary, Alberta. Since it wasn't that far away, I attended the conference. One of the keynote speakers was Dr Fred Provenza.

Dr Provenza presented some of his research to the group, and one of the tools he used in his research was molasses water. My ears immediately perked up because I could see a myriad of possibilities using molasses. I could hardly wait to get home to try it out.

Before I share examples of how I used molasses, let me explain the technical aspects of using molasses water for spot treatments. First off, you can get molasses in liquid or granular form. Only use liquid because granular will clog up your sprayer nozzles. Of course, I have never made that mistake, but I can certainly see how it could happen. I have tried two types of molasses: beet and cane. Make sure you stick with one or the other because once your animals are trained to one type, you will have to re-train them if you switch to the other type. Again, I have never made that mistake, but I can see how it could happen.

Before you start spraying the plants you want eradicated, your animals need to develop a liking for molasses. I do this by adding molasses to the mineral tubs. Sometimes I will spray the tub and surrounding area with molasses water and sometimes I will just dump molasses right into the tub. I'm not sure which method works best.

After three or four days of exposure to molasses, enough animals should be acquainted with molasses, and you can start spraying. Mix one part molasses to nine parts water. At this concentration there is enough molasses to attract the animals and the mixture still sprays nicely. Now it's time to start spraying.

I like to move the mineral tub near the area that I spray. That way, the animals are attracted to the area and are more likely to notice that the plants have been sprayed. Honestly, it's simple as that. One note though, this method may not lead to long-term affinity for the plants you sprayed, it will just clean out the area that you sprayed. For long-term effectiveness, I recommend using Kathy Voth's method.

Examples:

Fixing a crop failure: before I started using molasses water, I had the opportunity to test the effectiveness of strip grazing. It was the first year I worked for a neighbour setting up his grazing system. He had a 100 acre field that had been cropped the previous year and was under-seeded for hay. Nothing came up except weeds. Not just a few weeds, but a thick stand of weeds. Instead of spraying out the weeds and reseeding like he planned, I asked if I could strip graze the herd of 250 cow/calf pairs to see what would happen? He gave his okay, and I started giving the herd two day strips.

It took the cows about a half hour of walking around the first strip before they went to grazing. Once they got onto the weeds, they ate them down just like any other forage. I did two rotations of strip grazing that season. The following year, a lush, thick stand of alfalfa, timothy, and meadow brome came up providing a great crop of hay for the next three years.

I did the same thing one other time when a thick stand of volunteer canola came up instead of the intended under-seeded pasture mix. Again, it only took the cattle about thirty minutes before they started grazing.

Good-by to poplar saplings: molasses water on poplar trees can be a very effective method of controlling brush encroachment; it can be a lot of fun as well. We had some logging done in 2002 and, as expected, there was an infestation of suckers the next year. Those suckers kept growing rapidly and in 3 years they were close to twelve feet tall. Something had to be done, so I enlisted the help of Heather and Tiffany. I explained what I wanted done, filled up the backpack sprayer, and sent them on their way.

Tiffany and Heather loved spraying because once the cattle were onto the molasses water, they would follow the girls around and attack anything that was sprayed. The cattle stripped all the leaves off those saplings so that all that was left were bare sticks.

Heather and Tiffany liked brushing with molasses water so much they wanted to start a company where they went around teaching cattle how to eat poplar and willows. In fact, they were going to get a van, paint it up, and go to work making their fortune; they even had a funky slogan. Then we realized farmers would never pay to have a couple teenagers clear land for them.

Side note: after the trunks of those saplings dried out, the cattle broke them all off in 2-3 years, and to this day there are no poplars growing in that area.

Before   **After**

Common Tanzy: I have used molasses water on a wide variety of weeds and brush. The most controversial, though, is common tanzy. I know what you are thinking, tanzy is toxic, and you are correct. However, the tanzy that is toxic is ragwort tanzy not common tanzy. I know this to be true because I have been grazing tansy with cattle for at least fifteen years with no ill effects. I have also read a lot of research about tanzy toxicity.

My first experience with grazing tanzy happened by accident in 2004. That fall I had a group of 200 bred heifers locked into a drainage ditch that was loaded with weeds, particularly tanzy. Since it was late fall, all the weeds had gone to seed. Checking the heifers one day, I noticed that as they gathered around me, they were eating the seed bunches off the plants. In fact, there were nary any seed bunches left.

In our area, it is common knowledge that tanzy causes abortions if eaten; but luckily cattle don't eat tanzy. You can imagine my panic watching the heifers feast on the tanzy. When I got to the house, I started making calls to our Ag extension agent. He suggested I call another agent who had done a province wide survey of tanzy.

That agent told me that although ragwort tanzy is toxic, he did not find a single ragwort tanzy plant during his survey. Further, he assured me that common tanzy is very nutritious.

Since that time I have read a number of research papers describing tanzy toxicity. The effect of tanzy toxicity, documented in those papers, is that of severe liver damage. Interestingly, though, not a single paper documented that ragwort tanzy causes abortions. Each paper did note though, that many farmers & ranchers believe tanzy causes abortions even though there is no research to back up that claim.

The reason I have gone on about tanzy is because the myth that tanzy causes abortions has been so prevalent for so long that it is taken as fact. I still get challenged by ranchers, farmers, and weed inspectors every time I talk about grazing tanzy. The only ones who don't complain are my neighbour's cattle. In fact, his whole herd of 2,400 animals still eat tanzy regularly eighteen years after that first group of heifers sampled common tanzy for the first time.

Ground Preparation

There are times when I have wanted to create some type of ground disturbance. To this end, animals can be used to break up capped soil, trample old vegetation, establish seed to soil contact, or even smooth out eroded creek banks. To accomplish these goals, though, animals must be concentrated in the area to be disturbed or some sort of excitement must be created. Surprising to some, livestock are very versatile and when we want to create a disturbance, we are only limited by our imagination. Make no mistake, animals are capable of much more than we think and the cool part is that they work for food, not money.

A mineral tub is the easiest tool you can use to get animals to disturb an area. I have already mentioned a couple times where I have used a tub to get some herd impact. You can also use mineral blocks that are broken into pieces. If you want cattle to open up some paths in a dense stand of willows or buckbrush, smash a salt/mineral block into pieces and simply throw the pieces into the brush. I was introduced to this idea by our HM instructors who had cleared fifteen acres of buckbrush one summer on a community pasture in Idaho.

Following are several examples where I have used animals to modify the landscape using animals impact.

Examples:

Creek bank rehabilitation: a project I am quite proud of is how we reversed the effects of erosion on the creek that ran through our property. Our creek pasture had been continuously grazed for many years before we bought the place, thistle grew along the creek, and the banks were severely eroded as well.

One day, as I lamented about the condition of the creek, it dawned on me that plants can't grow on banks that are straight up and down. Plants require some slope so roots can get established. Thinking back to what I learned in the HM course, I wondered if we could use the cattle to smooth out the banks and maybe even trample the weeds that were growing along the creek. I decided to give it a try.

We had already fenced the cattle out of the creek with hi-tensile wire. Since the fence was built three paces off the bank, all we had to do was string poly wire at intervals across the creek. Over the course of three days, we moved the herd every 4-6 hours along the entire length of the creek. We did this once each year, and after two seasons, the results were pretty impressive.

By year three, the banks were covered in vegetation and were no longer getting eroded out during spring run-off. Even when we had a flood a few years later, the vegetative banks maintained integrity. Conversely, in dry years there is water the length of the creek, trapped in small pools.

I still build fences three paces off the banks because that gives cattle just enough room to move up and down the temporary paddocks, but not so much that they don't have to go into the creek and down the bank. I have been to ranches where the creek is fenced off, but the fences are right on the bank and it makes it very difficult to graze cattle when there is no room between the bank edge and the fence.

An added benefit to grazing your creek this way is that it keeps beaver from building dams. One ranch I managed had a major flooding problem because of beaver. The owner told me they never had a beaver problem until the creeks were fenced off. Further, after the fences were built, the cattle never went back into the creeks. Within several years, the centre of the ranch was so flooded there was no way to move cattle from the east side to the west side.

I think grazing a water course once a year mimics the effect of fire. In his book, *A Time For Burning*, Henry T Lewis explains that beaver do not inhabit an area for 5-7 years after a burn. After 7 years, though, the invasion is swift. It has been my experience that when a watercourse is grazed once a year, you don't have to contend with beavers flooding you out because the cattle keep brush from growing up and providing food for the beavers.

WATCH:
Creek Gazing

You can find this and other videos at simplyranching.ca/video.

Open up dense bush: I have had great results clearing under-brush simply by fencing bush areas separate from open pasture. I use bush as part of my drought insurance. To have that insurance, though, bush must be opened up so grass can grow in the under-story. I graze my bush areas at least once a year to keep woody species from growing and to freshen the grass.

Fire prevention: livestock can be a very effective tool to reduce the fire load around yard and farm sites. By getting your animals to graze these areas, they help to control weeds and reduce the ac-cumulation of old grass. Further, if you keep the grass vegetative, there is less chance of fire taking off and burning you out.

Bulls before breeding and your limpy pen are great to use for this task. You can also use these two small groups to prepare a site if you are planning a social gathering at your place. They will trample the grass so you can set up tables and chairs comfortably. Just make sure you do it 10-14 days before the event so the bugs have time to disperse the cow pats.

Getting muskeg to rot: often areas of muskeg can become pro-ductive pastures once the peat rots down. Of course, you have to first assess if the muskeg can hold up the weight of an animal. For me, yearlings work the best for this job because they are lighter than cows. The summer of 2008, I was able to reclaim forty acres of an old muskeg lake with a herd of 850 yearling steers.

The muskeg on that forty acres was still pretty spongy. I could see a wave go across the paddock when I jumped in one spot, but I was confident it could hold up a 700 pound steer. I held the herd in the forty acres for two days with a single strand of poly wire and step in posts. The steers made two rotations through the paddock that summer. The following year, the farmer who owned the land plowed up the forty acres and seeded tame grass. The soil was so firm he was able to use his big four-wheel-drive tractor!

That was the most successful muskeg I have ever reclaimed. I think it was so successful because there were no trees growing on the muskeg. That being said, I have also had a lot of success on forested muskeg.

Breaking up bedding packs: to break up bedding packs all you have to do is place a mineral tub on the pack. As the animals come to the mineral, they will loosen up the dense pack, which then allows seedlings to grow up though the hay or straw. The same process works to loosen up the last circle of dense material from bale grazing. In the case of bale grazing, though, usually all that is required are the smaller paddocks you should be using once your place is crossed fenced.

Keeping standing hay fresh: if you encounter a rainy period when you should be haying, taking your animals for a walk 3-4 times a day can be an effective way to freshen your standing forage. I have only done this once and had excellent results. It was at a ranch that relied heavily on pivot irrigation and for some reason it kept raining that year. The owner was getting quite stressed about his hay crop and was preparing to buy a bale wrapper. He asked me for my input because he didn't want to buy a wrapper and there were no custom operators in the area. I suggested walking the MOB around the hay fields to trample as much as we could so the hay would stay vegetative until we could hay. He was a bit sceptical, but after a few hours came back and agreed that it sounded like a pretty good idea.

For the next four days, I walked the MOB around the various hay fields four times a day. The animals were already trained to move as a group, so it was quite easy. Plus, I had fun practicing my handling skills. The animals didn't trample all the forage in each field. However, there was enough knocked down so when he was able to make hay there was plenty of new growth mixed with the mature plants and the feed was very good quality.

Cleaning old fence lines: on most places there are old fences that have been abandoned and have subsequently grown in with weeds and woody plants. Hogs are excellent at clearing fence lines. They are able to root up old wire and clean off all the grass etc. so all you have to do is roll up the wire. This is not a project for hogs that have a ring in the rooter, though, because it requires hogs that can root. One final note, since you will have to supplement while your hogs are clearing, you may as well use hogs that are less than 150lbs because you need to supplement them anyways. You may recall that when hogs are on high legume pasture they don't require any grain once they reach 150lbs.

WATCH:
Hogs Cleaning a Fence Line

You can find this and other videos at simplyranching.ca/video.

Packing snow or breaking capped soil: if you ever have a winter get together at your place and need a parking lot, a herd of cattle, strategically placed for a couple hours, will do the trick. By setting up a temporary wire around the area you want to use, then moving the cattle in for a couple hours, they will create a nicely packed area for people to park. I recommend doing this two days ahead of time so the cow pats have time to freeze. You can use the same technique if you have an area with capped soil. In that case, though, I like to spread a generous amount of seed before moving the cattle in. While milling about, the cattle will scarify the soil and walk the seed in.

Using Herd Mentality

When you have your animals trained to act as a group, they go through their daily motions as a group. Once you reach this state, the need to use electric fencing to concentrate your animals is greatly reduced. The animals will graze through a paddock together, not spread out all over the paddock. When one animal decides to try eating a new weed or nibble a new bush, there are numerous other animals that will join in. Once I came upon a small clump of poplar that was stripped of buds because of this phenomenon. Initially, I thought a herd of moose had gone through the paddock, except that moose don't herd.

Another interesting behaviour that occurs when you have sheep in the MOB is that calves like to stay close to the sheep. Those calves then learn to eat what the sheep are eating. That increases the palate of the calves and will transfer to the rest of your cattle via any replacement heifers you keep.

Livestock can be used as tools in a myriad of ways. The biggest hurdle, I find, is our willingness to try. I'm not suggesting that you can fix every landscaping issue with animals. What I am suggesting, though, is that before you spend money on equipment, labour, or chemicals, just try using your animals. You may be pleasantly surprised by what your livestock can accomplish.

WATCH:
Breaking Up Packed Snow

You can find this and other videos at simplyranching.ca/video.

CHAPTER 17

Should You Build It?

Before building infrastructure, I like to assess whether the fence or water improvement will pay for itself in one year. If you can recoup the expense of an endeavour in one year, you don't have to borrow any money. This was something Ranching for Profit taught, not as a rule, but as a way of assessing if the project should go ahead in the first place.

To demonstrate my point, let me relate a real-life example of how I evaluated a fencing project before construction. The property to be fenced was grazed by a group of breeding heifers twice during the growing season. It had a good perimeter fence, but no internal fencing. I estimated that if it was cross-fenced into seven paddocks, there would be about twenty days of stockpiled grass for the cow herd. I estimated the cost to fence at $11,000, including materials and labour.

The cost to bale graze that winter was $2 per cow per day. Since we had 800 dry cows, that meant each day we grazed stockpiled forage, we saved the ranch $1600. Therefore, twenty days of grazing was worth $32,000 (twenty days x $1600 per day) of bought feed. Subtracting the $11,000 cost to fence, that meant a total savings of $21,000.

We went ahead with the project because it was a significant saving in just the first year, and we would have the fences for years to come.

In the previous chapter, I discussed Stock Days per Acre (SDA). You may recall that SDAs are similar to bushels of grain per acre for a grain farmer. A grain farmer measures how many bushels are harvested from each field. The number of bushels is divided by the number of acres in the field to get bushels per acre. Let me repeat that the same can be done by a grazier. First, the herd size must be calculated in standard animal units (SAU). Then, multiply by the number of days grazed and divide by the number of acres in the paddock.

For example, let's say we had 615 1300-pound dry, pregnant cows that grazed from November 20 to 24 on 34 acres. To calculate SDs: 615 head x 1.23 SAU x 5 days = 3,782 SD. To calculate SDAs: 3,782 SD / 34 acres = 111 SDA.

Therefore, in this example, 111 SDA were harvested from the paddock. Now, a grain farmer would calculate the price per bushel when the grain is sold. Then they would figure out the revenue per acre. Graziers can do the same thing once they know the SDA and what one SD is worth.

During the growing season and up to October 31, I use the custom grazing rate for my area to give one SD a monetary value. I use custom grazing rates because it is the operator's choice whether they custom graze someone else's animals or graze their own. Starting November 1, I use the value of "normal" winter feeding to assign a monetary value. I use November 1 because I believe most people can graze until the end of October without much planning.

Here is an example:

- *Custom grazing rate in 2020: $1.25 per day per pair*
- *1 cow/calf pair (1300-pound cow) has a SAU of around 1.8 (from the HM table)*
- *Therefore, $1.25 per pair / 1.8 SAU = $0.69 per SD*
- *If we harvested 111 SDA from a paddock prior to November 1 with a cow/calf pair, the value of that grass would be: 111 SDA x $0.69 per SD = $76.59 per acre. If, however, we harvested 111 SDA after November 1 with a dry cow, the value of that grass would be an expression of how much hay it was replacing. In this example, it cost $2.10 per day per dry cow to bale graze. One 1300-pound dry pregnant cow has a value of 1.23 SAU. In this scenario, the value of one SD is: $2.10 / 1.23 SAU = $1.71 per SD.*
- *This means that 111 SDA of grass after November 1 is worth: $1.71 per SD x 111 SDA = $190 per acre.*

Here are the actual results from the property I mentioned earlier that we decided to fence:

The property was around 520 acres, half open, half dense willows with some large poplar trees. As I stated earlier, I estimated that there would be twenty days of grazing for 800 cows. Once the property was fenced, we moved 615 dry cows into the first paddock. We added cows and took some away, finishing with 802 head on January 1.

We harvested 125 to 175 SDA in each of the first six paddocks. Grazing in the last paddock was cut short by fifteen days because we had set out too many bales with net wrap removed. We were concerned there would be excessive spoilage if it was left out in the field unprotected until the next winter, so we switched to bale grazing January 2.

The dry herd of cows began grazing stockpiled grass on the new property on November 20 and finished on January 1 for a total of 41 days. My estimation was way off! However, in my defence, I have become very cautious when estimating grazing days during the winter because grazing through snow and blizzards can end winter grazing abruptly.

Over the course of the 41 days, 43,376 SDs of grass were harvested. As I pointed out, we would have been bale grazing if not grazing. Therefore, the value of each SD was worth $1.71. In the end, we harvested $74,173 ($1.71 x 43,376 SD) worth of grass from that property. Our cost to make this happen was actually less than $9,000 (fencing materials, labour, and winter-water system). I think that is pretty cool!

Another way to look at this scenario is to consider the cost per day to feed each cow. The expense for this property was around $22,000 (rent, fencing, and water development). I know the fencing and water costs should be amortized, but when we first started, every improvement had to be paid for with current revenue, not borrowed money. With the growing season included, we harvested 49,945 SD.

Remember: 1 SD is equivalent to 24 lbs of dry matter.

To calculate the expense or $ per SD: $22,000 / 49,945 SD = $0.44 per SD. To find the cost to feed a cow for a day (November 20 to January 1): $0.44 per SD x 1.23 SAU = $0.54 per day per cow. Add in yardage (labour, truck, and quad): $0.54 + $0.05 = $0.59 per day per cow.

To calculate how much money grazing stockpiled grass saved:

- *$2.10 per day (bale grazing) x 41 days = $86.10 per cow*
- *Subtract $0.59 per day (stockpiled grass) x 41 days = $24.19 per cow*
- *Total savings = $61.91 per cow*

At the beginning of this section, I wrote about economic sustainability. Decreasing winter feed cost is not the whole story to achieving this mission, but it is certainly a significant portion.

CHAPTER 18

Building Your Room for Success

Some time ago I listened to an interview with author David Weinberger on CBC Radio. The host was asking questions about Weinberger's book *Too Big to Know*. Weinberger posed the question, "Who is the smartest person in the room?" His answer... the room!

In the book, the author talked about how the internet and our access to the internet have radically changed the face of knowledge and expertise. In our current environment, anyone can be an expert because of the availability of information. This is to say, in any room there may be a host of experts on a variety of topics. It got me thinking about my "room" of experts.

Since I have had a few careers in my life and have met some extraordinary people, I have a healthy list of experts I can call for advice or free information. The people I trust most are people who have actually tried something. I stay away from people who can tell me all the reasons why something won't work when they have never actually stepped up.

The most powerful advice comes from people who are intelligent, have education, and who have a lot of experience. I strive to have people like this in my 'room'! Not only do they give me a great deal of help, they are also very interesting to talk with. They tend to have a lot of passion and a wealth of knowledge on many more topics than the one at hand.

Having a group of experts I can call anytime has saved my bacon more times than I would like to remember. In my ranching life, I have my friend Glenda, who coached us along as we learned how to graze hogs. It was amazing how many people who had never raised hogs outside of a pen knew why it couldn't be done. If it hadn't been for Glenda, I don't think we would have had the courage to try. Things worked out very well and now hogs are my favourite animal to graze. And Glenda has been a good friend for over twenty years!

As I got to know more people in the Ag community, my room expanded. It wasn't a fast expansion because our practices were outside the realm of normal and we were often referred to as the "hippies who live on Jackson's Hill."

A big break for getting to know more agriculture people came when a neighbour—Kelly, the largest cattle producer in the county—asked me to set up electric fencing for him. He was having trouble finding the know-how and the time to set things up himself. I was blindly confident I was up to the task. In fact, I thought it was my dream job!

That first year was hell! I kept asking, "What have I gotten myself into?" Luckily, I am pretty stubborn, so I tucked my chin and worked like hell to make it happen. I knew there were a lot of eyes on me.

Once Kelly saw how effective electric fencing was and his wife wasn't constantly answering phone calls about cattle being out, things became much easier. Kelly and I became friends, and he is one of the people in my room. He has helped me out more than once.

Kelly was there for me when I overestimated how many cattle we could run on a new rental property. Half way through the first rotation I knew what I had done, but I did not know what to do about the situation. I started to panic and after one sleepless night, I called Kelly to ask what I should do.

First, he said, "Don't worry. There are a lot worse things that can happen." Then we went for a drive. Kelly showed me some land he wasn't really using that would be easy to electric fence. Then he suggested I send the whole herd to a property he was involved with. Two weeks later, the cattle were on the trucks to the new property, and we ended up running cattle there for two years. I relate these stories because you never know how things will turn out when you ask for help. The critical part is that you have to ask.

A couple months ago I was explaining to a friend about my room. He wasn't quite getting it. He feels he has to get an outside expert to help him with business ideas. I firmly pointed out that he was wrong. He already had access to all the help he required among the people he knew.

Now, Bill is one of those guys that everyone likes. In fact, I don't know anyone who does not like him. I envy that! Not only is he likable, he is also incredibly handy at many things. Again, I envy that. People naturally gravitate to him and want to help him in any way possible. However, Bill does not ask! For some reason, he feels he is a bother to people if he asks. "How stupid!" I told him. People love helping. I know I have felt very good the times I have forced my advice on him.

Once I talked some sense into him, he realized he does have a large room of experts. He just has to ask. In fact, a month after that conversation, his business ideas began happening in a big way. He just had to reach out to one of his contacts.

It has been my experience that we never really know how things will play out once we start asking for advice from our room of experts. A key point, though, is to build your room wisely. Like I said earlier, the best advice comes from people who are intelligent, educated, and who have tried things. Stay away from contrarians. These are the people who reject every new idea or anything out of the ordinary. These are the people who sit in the coffee shop every morning and complain about everything!

If you are not sure who these people are, just ask yourself how you feel after being around a person. If you are not uplifted and energized, then that is a person to avoid. About seventeen years ago, I went to the coffee shop because I thought I was missing out. My friend Kelly was not happy because he wondered who they were going to talk about if I was there!

I went to coffee for two days that week. Each time I left the coffee shop, I went home feeling deflated and doubting my abilities and plans. Stay away from people who make you feel that way.

My final thoughts on building your room:

Ask questions of people you meet. You never know who may be an expert, and people who are experts are usually pretty interesting!

- Get enough info so you can contact them at a later date. You never know when their expertise could help you out. I don't always ask for contact info directly, but I remember details like where they live, mutual acquaintances, where they work, etc. It may take some detective work if you ever want to pick their brain, but at least you will have a place to start.

- *ASK!* The only way to get the help you require is to ask. It took me several years to be comfortable calling up someone I didn't know very well. Each time I felt trepidation picking up the phone, I would remind myself, though, that people feel honoured to be asked for their advice.

- Park your ego. If you don't know or don't understand something, ask for clarification rather than keeping your mouth shut. Most people are pretty tickled when they can tell you something new.

- If you do not have a personal connection with someone, they are not in your room. Sending an email to an author or speaker is not the same as emailing someone you met at a party. People in my room know my name.

That's all I have to say about that. Good luck and get started building your room of experts today!

From the Tom Krawiec Dictionary, 2021 Edition

Contrarian: a person who automatically takes an opposing view to anything you propose. It is a habit formed by upbringing or the crowd they hang with. Contrarians are very dangerous creatures and should be avoided like the plague. They like to criticize your dreams, your plans, and your life in general.

Don't just speculate, Googleate: the act of looking for factual information rather than relying on the accuracy of what someone is saying. It should be used during a coffee shop conversation, but rarely is.

CHAPTER 19

Training Employees

In 2004, I stumbled upon a book by Robert Kiyosaki entitled *Rich Dad's Cashflow Quadrant*. It really changed the way I looked at what I was doing. I thought we were building a great business, but after reading Kiyosaki's book, I realized we were just creating a job for ourselves. The Ranching for Profit course also taught that we do not really have a business until our first employee is hired.

Once I changed my thinking and realized that to build our operation we required employees, our operation began to grow quickly. When someone else was helping with the day-to-day chores, I had time to find more grass to lease and more cattle to graze. And my wife was able to go to college and fulfill her dream of becoming a nurse. At 42 years of age, it was a pretty courageous undertaking!

When hiring an employee, you require another set of skills, and those skills don't involve livestock or grass. You must acquire the skills to manage and train people if you are going to be successful. Luckily for me, I had a great teacher in my first rig manager, Duane Carol. Under his tutelage, I figured out how to train and inspire the people I hire.

One day, a hired hand asked how she should set up the gates in the corral as we prepared to vaccinate a herd of replacement heifers. In response, I asked, "What do you think?" This began a discussion—or really, a lament from her—that she needed to start thinking for her-self and using the tools she was learning, to make decisions.

I assured her that most young people have that same experience. Getting people to think for themselves was one of the hardest things I had to do when training new guys on the rigs. It hasn't changed much since I had my first crew over twenty years ago! What has changed, though, is the approach I now take to training people.

Duane, my first rig manager, had a saying: "Teach the guy below you to do your job so you can learn the job above you." He tried to instil that thinking in all his employees and was pretty successful. He developed a lot of drillers and rig managers over the course of his career. The cool thing about his method is that it built continuity into the system. If someone was missing for some reason, there was another person trained and ready to fill that position.

In 2010, I read General Rick Hillier's book *Leadership*. I used it to develop my training program even further. General Hillier wrote about building continuity into the system the same way Duane professed, "Train the guy below you." General Hillier also wrote about keeping your eyes open for the special person who shows up every now and again.

Once you identify that individual, put most of your energy into that person because they will become your superstar. Even further, Hillier believes that if you put six of these "superstars" together, they can lead a battalion of a thousand soldiers!

From Duane's and General Hillier's ideas, I developed the "Tell Me. Show Me. Watch Me." method of training.

The first step, "Tell Me," involves explaining the how and why as well as the dos and don'ts. I find the "why" is very important because it gives the new person the context of the task in relation to the bigger operation.

Step two, "Show Me," simply demonstrates how to perform the task. There are some tasks that must be done a certain way for safety reasons, while other ones are just my preference. For the latter tasks, I make a point to let them know that this is the way I do it. Once they are comfortable with the task, they may figure out a way that works better for them. To me, the result is more important than how it was accomplished.

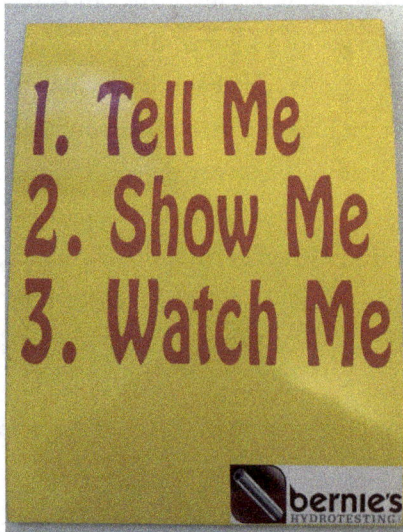

1. Tell Me
2. Show Me
3. Watch Me

bernie's
HYDROTESTING

Finally, the "Watch Me" step. Here I just shut up and observe. It can be a very difficult thing to do because we have a strong desire to jump in and give advice when someone is making a mistake. The goal is to get the new person thinking and making decisions. When that is the goal, they need to make mistakes, identify those mistakes, and correct them. It doesn't work if someone else is doing all the correcting! It really helped me remember to keep my mouth shut once I had signs made up that boldly stated, "Tell Me. Show Me. Watch Me."

There are probably a number of people reading this asking, "what happens if the trainee is making a catastrophic mistake?" Then of course, you need to step in, but really, how often does that happen? Not very!

Not every person you hire will become a superstar. Some are not really engaged. Some are just lazy. That's fine, and it doesn't make them bad people; it just means that they are not what I am looking for. The sooner I find that out, the sooner I can make room for someone I do want on the team. As my Polish grandfather would say, "You can't make a silk purse out of a sow's ear!" so I like to start with silk. That being said, sometimes we miss someone with great potential because we are not being a good boss. In my experience, there are as many poor bosses as there are poor employees.

As a young person is developing, they may become nervous and maybe even scared of tasks I give them. I understand being scared because I still do things that scare the heck out of me… just ask my horses!

When an employee is feeling less than confident, I reassure them that I won't put them in a situation that would be catastrophic if they mess up. Mistakes are part of life and learning.

They should also remember that they have the tools to figure out what to do when things go south. Each step we take, each hurdle we overcome, we become more competent as people and as workers.

I get a thrill out of seeing young people grow and develop their expertise, confidence, and independence. It warms the cockles of my heart when a young person's talents start to shine through. I know I have done a good job when a young person corrects me after I say or do something stupid. My pride may be hurt a little, but that heals pretty quickly!

Finally, I know there is always talk about the poor work ethic of the new generation. For some, that may be true; I know we have hired our share. However, there are a lot of highly motivated, conscientious, hard-working young people looking for an opportunity.

Before I finish this topic, let me share my thinking on the difference between an intern and an apprentice. To me, an intern is someone who has limited skills but has a keen interest in ranching. These are people who must be trained from the ground up. In my experience, these recruits don't really contribute to an operation in a meaningful way for at least three to four months. In my opinion, they should expect a wage commensurate with their contribution to the ranch.

An apprentice, on the other hand, is a much different beast. An apprentice is someone who comes with a lot of skills. The first official apprentice I hired graduated from the Sustainable Ranching program at Thompson Rivers University in Williams Lake, British Columbia. She worked as an intern for several different operations, including a large ranch in New Mexico. She had also worked on a dairy farm in New Zealand for nine months. On top of that, she completed an animal handling course with Curt Pate.

With her skill set, she was able to take over a lot of duties within two weeks. Because of her passion and prior experience, she contributed to our planning meetings in a meaningful way within a month.

Now, Zetteh was an exceptional find. Out of forty-five applicants for the position, she was the only qualified candidate. To find and keep someone like Zetteh, you have to pay them much better than an intern. If your operation can't afford to pay an apprentice a living salary, then maybe it isn't ready for one. Certainly, though, don't expect someone capable of being an apprentice to stick around when they can't earn a living wage.

In my mind, an apprentice should be someone you are training to take your position. That means they have the authority to make decisions, plan grazing rotations, make proposals for expansion, and train the new hands coming on. If done right, and with intention, an apprentice should be ready to take over your job within three to five years. If at that point they decide to go somewhere else, they will have trained another person ready to take their place.

Being a boss takes a whole new skill set and many people are not suited for the job. There is nothing saying you have to become a boss. It is good, though, to recognize that fact as soon as you can. It will make your life and the lives of those around you much better.

Conclusion

It is easy to make ranching complicated. I have mentioned numerous times in this book that ranching should be simple, easy, and fun. If it isn't we are doing something wrong. I have shared my thoughts on many aspects of ranching and how to simplify your operation. You may have noticed, though, that I have not talked much about how to increase your revenue. The reason is that this is not my strong suit.

I am not an expert on marketing which is why I have not written about it. Most of what I have shared are ways to decrease your expenses. Decreasing expenses is just as important as increasing revenue, but is not the only part to creating an economically sustainable ranch. Fortunately, there are other people who are very good at finding ways to increase revenue. I urge you to seek out these people and learn from them. It is my belief though, that the first 2-3 years should be spent lowering your expenses before looking at improving revenue streams.

I see many people developing animal genetics or direct marketing meat before improving their pastures. These can be great endeavours, but your grazing has to be in shape first. A successful ranching system has to start with the grass, period.

Producing high quality, high-volume grass is the foundation of all the other projects. It will give you flexibility and a firm base that will carry you when things get tough. I have shared numerous examples where great grass helped me maneuver when market conditions changed, but it is pretty difficult to pivot if you don't have many options. When your land is producing a high volume of grass and your management keeps it in a state of high quality, you can adjust to changing environmental, economic, or political factors much better than when you don't have lots of high quality grass.

This book has provided a lot of information about how I ranch. I'm certainly not the only game in town and I realize there are many other ways of performing the day-to-day tasks of ranching. I wrote this book to get you thinking about how to make your operation simple, easy, and fun. I am a bit embarrassed to admit, though, that this book can be summarized with one question. It is the question I have asked myself for almost twenty years and continue to ask myself on a daily basis.... "CAN A 12 YEAR OLD DO THIS?" I encourage you to start asking yourself the same question with every task you perform and every dollar you spend.

12-Year-Old Ranching Hacks

H*ere are a few more ideas to make ranching easier. As you start to apply my philosophy of ranching, you will certainly discover more hacks of your own.*

Train your animals to come when called

This will make your life much easier, particularly when you want to get animals out of a bush paddock.

I recommend you use something other than your voice so anyone can gather your animals. That way, it doesn't matter who goes out to move the group. I have tried bells, banging on a pail, and banging on a stock-trough when pelleting calves that were still on the cow. I have switched to using whistles because they are easy to carry around.

WATCH:
Calling 300 Heifers with a Whistle

You can find this and other videos at simplyranching.ca/video.

Pick and sort cull cows overnight

If you want to get rid of some late calvers or cows with poor calves, sort off the runty calves at vaccinating time. Put them in a pen. Kick the rest of the herd out to pasture. Make sure there is an easy path from the pasture to the pen. Go to the house. In the morning, the mothers of the penned calves will be laying down outside the pen. Just close the gate and you have your cull cows. This technique can be used to make a variety of sorts; it just requires a little thought and planning.

We have used this method many times. The most extreme time this method helped us out was when a group of two hundred summer calving pairs got in with four hundred spring calving pairs. We brought the whole group in and sorted calves from cows. Since the summer calves didn't have ear tags, we then sorted off all the calves with no tag. We put those calves in a pen and sent the rest out to pasture. The next morning, all the summer calving cows were in the alley chewing their cud and our job was done.

Sort in a pen or out on pasture with poly wire.

When your animals are trained to avoid electric fence, they fear the poly wire. To use this method of sorting, all that is required is two people, one on each end of the poly wire. Use the poly wire to push the animals you want toward the gate and lift the wire over the ones you don't. Our kids were ten when they started helping us sort in this manner. Cowboys don't like doing it this way, but it is usually much quieter and faster than the cowboy way!

Calve on grass

Although this is one of the tenets of both Holistic Management and Ranching for Profit, it also enables my mature friends to continue ranching. In our latitude, I prefer May 15 over May 1 because there is still a good chance of getting a cold blizzard in the first ten days of May. If there is a blizzard during second part of May, it tends not to be as cold and nasty.

Calve on fresh ground

I have been paranoid about having a scours outbreak. That is the main reason the herd gets moved every two to three days during calving. Of course, the other reason is to get the grass clipped. It has been my experience that if newborns stay away from areas fouled by mud and manure; they stay healthier.

Set up one or two blizzard sites

A blizzard site is a spot animals can go to get out of the wind. Normally, that would be a place protected by trees. If you don't have such a place, several free-standing wind breaks work well. On the protected side, place enough hay bales for three days and several straw bales. If a blizzard is imminent, cut the twines on the bales and move the herd to the bales. It is far easier than scrambling to put out feed and bedding as the blizzard is bearing down.

The first time we used a blizzard site was in a blizzard on May 8, 2006. We were still keeping cows all year, and the owner would not let us calve later than April 25. I didn't think the incoming weather was anything to worry about. However, when I went out in the morning, boy was I kicking myself! There was a strong cold wind blowing, and we got two inches of snow overnight—those poor calves!

I called my wife and the summer student and between the three of us, we were able to walk the herd to the blizzard site. In forty-five minutes, it was a different scene. The cows were munching away on the new hay and the calves were cavorting about in the fresh straw. I reminded myself to take preventive action in the future and move the animals before the blizzard, not during!

Stop tagging

If you are a purebred operation, then tagging is a requirement. However, most people are commercial producers thinking they need to do what purebred people do. If you want to know the dam of a particular calf, all you have to do is pen it up for a few hours and the mother will show up. If you send small groups to different pastures, do the same thing. Pen the required number of calves for a couple hours and the mothers will show up.

My friend Marty does a version of this when he brands. He sorts off all the calves, then moves the cows back out to grass. Invariably, the cows with calves come back to the corrals. The ones still out on pasture are his open or late cows. He just gathers them up and moves them to their own pasture for future consideration.

Whether you agree or disagree, one thing we can all agree on is that tagging is dangerous work, and it is usually being performed at a critical work time. When someone is injured tagging, they are laid up for the rest of calving and for other critical duties like seeding or spring field work. I find it interesting to hear stories of someone getting run over by a cow and laid up for three to six weeks. The rest of the calves did not get tagged or given shots and they somehow survived and did as well as the ones that did receive all the human attention!

Author Website QR Code
https://simplyranching.ca/

PM Store Author's QR Code
https://pagemasterpublishing.ca/by/tom-krawiec/

To order more copies of this book, find books by other
Canadian authors, or make inquiries about publishing your
own book, contact PageMaster at:

PageMaster Publication Services Inc.
11340-120 Street, Edmonton, AB T5G 0W5
books@pagemaster.ca
780-425-9303

catalogue and e-commerce store
PageMasterPublishing.ca/Shop

About the Author

Tom was raised in the oil patch in Northern Alberta, but after spending summers with relatives who farmed and ranched he realized that's where his heart was. Tom formally entered the world of ranching in 2000 near Athabasca, Alberta, Canada by custom grazing 40 cow/calf pairs on their 380 acre farm.

His education includes Holistic Resource Management, Bud Williams Low Stress Animal Handling, and Ranching for Profit programs, and everything grazing related in between. Tom started grazing multiple species at his Athabasca ranch and it was a common sight to drive by and see hogs, horses, sheep, cattle and turkeys grazing in paddocks together. The learning curve was steep and sometimes very humorous. This is where he learned important lessons such as roping hogs to load them is not a good idea and that 200 turkeys leave an impressive herd impact on your deck. He eventually expanded his custom grazing operation to 5500 acres of leased land and 3000 yearlings.

Tom now manages ranches for others and has grazed in droughts as well as flooding. His constant quest for more knowledge of grazing and grass management continues and he lives by the adage "any day in the saddle is a good day".